*Divorce: An American Tradition*

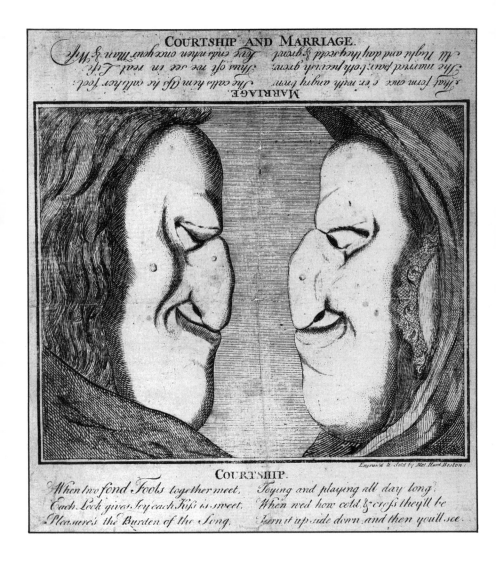

"Courtship and Marriage," an eighteenth-century illustration by Nathaniel Hurd. When turned upside down, a smiling courting couple becomes a frowning married couple. Courtesy American Antiquarian Society.

# DIVORCE

## An American Tradition

GLENDA RILEY

*New York   Oxford*
**OXFORD UNIVERSITY PRESS**
*1991*

## Oxford University Press

Oxford   New York   Toronto
Delhi   Bombay   Calcutta   Madras   Karachi
Petaling Jaya   Singapore   Hong Kong   Tokyo
Nairobi   Dar es Salaam   Cape Town
Melbourne   Auckland

and associated companies in
Berlin   Ibadan

## Copyright © 1991 by Glenda Riley

Published by Oxford University Press, Inc.,
200 Madison Avenue, New York, New York 10016

Library of Congress Cataloging-in-Publication Data
Riley, Glenda, 1938–
Divorce—an American tradition / Glenda Riley.
p.  cm.   Includes bibliographical references and index.
ISBN 0-19-506123-3
1. Divorce—United States—History.   I. Title.
HQ834.R55   1991
306.89'0973—dc20   90-47746

9 8 7 6 5 4 3 2 1

Printed in the United States of America
on acid-free paper

*For My Mother*

# Preface

For nearly twenty years, I have been fascinated by the topic of divorce. As I explored the history of nineteenth-century American women, numerous cases of distressed wives and divorced women piqued my curiosity. The ranks of western American women were especially replete with distraught and deserting wives, female divorce-seekers, and divorced women. Determined to understand these women's circumstances, I began to try to reconstruct the historical evolution of divorce from American colonial days to the present.

My quest for information was all the more interesting to me because I am divorced, as are many of my friends and acquaintances. My questions about divorce, and theirs, occasionally made my search almost compelling. I was soon convinced that establishing a historical context for divorce would not only make it more comprehensible to researchers, but also to contemporary Americans who experience it, shape it through law and policy, and attempt to help family members, friends, and clients survive it.

Of course, researching the history of divorce forced me to clarify my feelings about divorce. I have come to believe that divorce is a remedy for mismatches. Too many people chose a life partner under some form of duress: youth, romantic illusions, lack of self-awareness, a tendency to present only one's best side during courtship, sexual attraction, parental and societal pressures, a need for psychological or financial support, a desire to have children, a fear of being alone, and advancing age, among others. Most of these pressures to marry are understandable, while others are healthy desires. But pressures to marry must be balanced by an understanding of oneself, one's potential mate, and the nature of marriage. When these factors go unbalanced, the result is too often un unsatisfying or destructive marriage—a mismatch.

Divorce releases people from a lifetime of living with unsound judgments regarding their potential mates. In addition, it removes others from debilitating situations that develop after marriage; situations that are sometimes difficult to foresee, including spouse abuse or sexual exploitation of children. In yet other cases, husbands and wives discover that they are unable to function in a heterosexual marriage because of their own homosexual orientation.

My support of divorce is not meant to suggest that we abandon our ideal of marriage as a lifetime commitment. It simply asks that we recognize that many couples are unable to sustain such a commitment. Divorce allows these couples to dissolve dysfunctional unions. After divorcing, they have the opportunity to reassess themselves and perhaps form functional, satisfying marriages with other mates.

Agreed, some people may exploit lenient divorce laws by divorcing their spouses heedlessly and impetuously. But the historical record indicates that people forced to stay in failed marriages against their wills usually sabotage their relationships, either consciously or unconsciously. Adultery, abuse, and desertion are frequent occurrences among such couples. In addition, other distraught spouses find illicit ways around restrictive divorce laws. They falsify grounds for annulments or perhaps leave the United States to obtain a divorce elsewhere. When people remain in difficult marriages or circumvent the law, it often costs them, their spouses, children, relatives, friends, and employers, as well as society, dearly.

These ideas, which are not original with me, come up many times throughout the following narrative. I also present other viewpoints concerning divorce fully and, I hope, fairly. I would encourage readers who disagree with my position from the outset to persevere, both to give my view a hearing and for what they might learn about the historical development of divorce in the United States.

My major goal, however, is not to argue for the rightness of my view of divorce and the wrongness of others. Rather, my primary concern is to reveal that the historical conflict between anti-divorce and pro-divorce factions has prevented the development of effective, beneficial divorce laws, procedures, and policies. Today we still lack processes that move spouses out of unworkable marriages in a constructive fashion and get them back into the mainstream of life in a stable, productive condition.

I maintain that we need to reshape the institution of divorce with an eye to the people involved rather than according to our own attitudes toward divorce. History demonstrates that people will divorce. Given this reality, it serves everyone's best interests—even those people who oppose divorce—if the divorce process helps divorcing men and women become healthy, effective individuals. For instance, divorced people who live in poverty cannot adequately care for themselves and their children. They work little, often in dead-end jobs, and, frequently cannot provide their children with vocational training or higher education. Society as a whole suffers from this underutilization of resources. Yet no-fault divorce provisions, combined with a lack of sufficient job retraining and child care, cause many divorced Americans to remain unemployed. A promising future remains out of their reach and their children's.

If we draw upon history to see that conflict over divorce in the past has led to problem-laden divorce, we can create more effective divorce in the future. We can continue to see lifetime marriage as a worthy ideal, while providing salutary divorce for those who are unable to achieve the ideal.

To understand how and why contemporary American divorce needs to be rethought and reformed, we need to begin with a colony-by-colony examination of American laws, procedures and attitudes for or against divorce. If we are to understand later state and regional patterns, it is necessary to understand the status and development of divorce in each of the American colonies.

Throughout, I discuss divorce only in the United States. The use of the term "America" in the title of this book indicates that the study encompasses the American colonies before 1776 as well as the United States after its declaration of independence from England. The study touches only lightly on African American, American Indian, and other specific groups. Research into the history of divorce among these peoples is just being undertaken.

I also refer to the "institution" of divorce. Although some social science definitions oppose this usage, most support it. The *Encyclopedia of Anthropology* defines an institution this way: "All known human societies have standard ways of doing things which consist of . . . *norms* serving as goals . . . *roles* constituted by norms, and *patterned behavior* attached to the norms and roles." It adds that "the whole system of standardization of a behavioral pattern is called an institution." Other experts define an institution as "a pattern of behavior that focuses on a central theme" and "practices based on similar principles that display some degree of regularity."[1] My usage of the term fits within these meanings.

In researching and writing this study, I have incurred many debts. The staff of the Henry E. Huntington Library in San Marino, California, was extremely helpful and cooperative during my tenure as a research fellow in 1988–89. I owe special thanks to Peter Blodgett, Doris Smedes, Virginia Renner, Elsa Sink, Leona Schonfield, and Mary Wright.

Staffs of other archives and libraries who deserve thanks are the Center for Western Studies, Augustana College in Sioux Falls, South

Dakota, especially curator/managing editor Harry F. Thompson; Cass County Clerk of Courts Office in Fargo, North Dakota; Fargo Public Library; University of Iowa Law Library in Iowa City; Linn County Clerk of Courts Office in Cedar Rapids, Iowa; Logan County Clerk of Courts Office in Guthrie, Oklahoma; Louisiana Historical Center at the Louisiana State Museum in New Orleans, especially librarian Kathryn Page; Minnehaha Clerk of Courts Office in Sioux Falls; Historic New Orleans Collection; North Dakota Legislative Council in Bismarck, especially librarian Marilyn Guttromson; North Dakota Institute for Regional Studies at North Dakota State University in Fargo, especially director John E. Bye; State Historical Society of North Dakota in Bismarck, especially historian/editor Larry Remele; University of Northern Iowa Library in Cedar Falls; Special Collections, Northeastern State University Library in Tahlequah, Oklahoma, especially Delores Sumner; Oklahoma State Museum and Historical Society in Oklahoma City; Archives and Manuscripts Division of the Oklahoma State Library in Oklahoma City; Oklahoma Territorial Museum in Guthrie, especially Wayne Ward, Director, and Gordon Moore, Curator; Sioux Falls Public Library; Siouxland Heritage Museum in Sioux Falls, especially collections assistant William Hoskins; Armistad Research Center at Tulane University in New Orleans; Special Collections, Howard-Tilton Memorial Library at Tulane University; United States Court of Appeals Law Library in Pasadena, California; and Virginia State Library in Richmond, especially reference librarian Sarah Huggins.

Both the research and writing phases of the project were aided by generous research grants from the Henry E. Huntington Library and the University of Northern Iowa in Cedar Falls, for which I am very grateful.

In addition, colleagues and friends helped the writing phase by selflessly reading, and rereading, drafts of the manuscript as it evolved. Martin Ridge of the Huntington Library gave generously of his time, support, and suggestions. Annette Atkins at St. John's College in Minnesota offered incisive and extremely helpful suggestions. And John Johnson at the University of Northern Iowa provided regular, on-the-spot help and encouragement that smoothed out rough spots in the manuscript on more than one occasion.

I would also like to express my appreciation to others who read and offered helpful comments on sections of the manuscript: Norma Basch, Robert L. Griswold, Kermit L. Hall, Sandra L. Myres, Mary Beth Norton, Susan C. Peterson, and John Phillip Reid. I am grateful for the ideas and information contributed by Father Thomas E. Buckley, James G. Chadney, Stephen Cox, Jeffrey J. Crow, Carl J. Eckberg, Lillian Gates, Thomas L. Hedglen, Karen Lystra, Howard Shorr, Susan

Kallmann Puz, Harold B. Wohl, Carole Shelley Yates, and Paul M. Zall. At Oxford University Press, Sheldon Meyer provided enthusiasm and suggestions, while Leona Capeless raised questions and lent grace to the style. Of course, I remain responsible for any errors of fact and for my opinions concerning divorce and modifications in it.

I am also indebted to the many divorced women and men who contributed their time and stories. Because they provided a large amount of easily collected interview material, I did not have to identify a target sample or follow complicated procedures to collect information. I usually conducted interviews as they came to hand; at conferences, in stores and restaurants, on airplanes, and at club and organization meetings. I simply had to mention that I was writing a book on divorce and more people volunteered to be interviewed than I could accommodate. "Do you want another story?" they asked, or "do you need more evidence for your study?" They seemed to have an excellent recall of the details of their divorces, and, in some cases, their eyes misted over although the events had transpired years ago.

Often, these people were loath to let me go, for in entrusting me with part of their personal histories, they also seemed to have temporarily bonded with me. They frequently wanted to exchange names and addresses. They occasionally expressed their wish that their stories would help other people. One Charlottesville, Virginia, man hoped that my book, and his tale should I choose to include it, would help others marry more wisely than he did initially. "You can't change a person," was his advice. "You better like him or her as is at the time of marriage."

I have, of course, changed or omitted the names of these willing interviewees in the study that follows. Like them, I hope that their stories, and the larger story of divorce in America, helps others.

*Milwaukee*                                                                                     G.R.
*November 1990*

# Contents

*Divorce: An American Tradition*

# Introduction
# An American
# Tradition

⊐ During the past several decades, people all over the world have expressed astonishment and disbelief concerning the spread of divorce in the United States. Americans in particular have examined divorce from every angle, often reproaching themselves and their tension-laden, urban, industrial society for making divorce a widespread American phenomenon. The historical record, however, indicates that contemporary American divorce is more than a recent outgrowth of a troubled modern society. American divorce has a long and venerable history: Puritan settlers first introduced it in the American colonies during the early 1600s. The resulting institution of American divorce was vital, and growing, long before late twentieth-century Americans carried it to its current state.

Still, of the more than one hundred and fifty Americans whom I interviewed concerning their knowledge of historical divorce in the United States, the majority were surprised to learn that Puritan colonists condoned and practiced divorce. I concluded that as contemporary Americans grapple with the complexities of life and personal relationships in the late twentieth century, they tend to look back at early America as an idyllic age characterized by simplicity and family harmony. They firmly believe that enduring marriages and peaceful family life were the rule among colonial Americans.[1]

While this image of the "good old days" of the nation's marital harmony may be reassuring, it is inaccurate. We see our ancestors as we wish they were—peaceful and loving—rather than as they really were—human and often contentious. Surviving court and other records reveal that a number of colonial Americans sought divorces after they experienced disillusionment or dissatisfaction in marriages. The first American couple to divorce obtained their decree in 1639 from a Puritan court in Massachusetts. Anecdotal evidence indicates that

3

untold numbers of other colonists simply deserted their unwanted or offending mates.

Thus, divorce has been developing and growing in what is today the United States for over three hundred and fifty years. Today it is a customary, or traditional, way to resolve marital incompatibility. Yet, despite its increasing prevalence, only two historians have traced its development in American society from the early 1600s on; one in 1904 and the other in 1962.[2] It is the purpose of this book to provide a historical overview of divorce from its beginnings to the present so that today's divorce-prone Americans can understand how divorce became woven into the fabric of their society.

This is not to suggest that all Americans accepted divorce. Many opposed divorce in the past, and many continue to oppose it today. Over the years, critics and opponents of divorce have maintained that marriage is a religious sacrament and a lifetime undertaking. In their eyes, the growth of divorce signaled impending breakdown and disintegration of the American family.

Opponents of divorce usually believed that marriages should be terminated only for the reason stated in the Bible: adultery. As a result, some supported restrictive divorce statutes stipulating only adultery as a ground for divorce, while others were willing to accept other limited grounds as causes for divorce, such as consanguinity and insanity. Although critics of divorce usually condoned the dissolution of marriage by annulment, not all thought that divorce of bed and board—a limited divorce that prohibited remarriage—was valid. They also strongly opposed migratory divorce, in which divorce-seekers fled strict laws in their own home jurisdictions to obtain divorces in more permissive states, territories, or countries.

On the other side of the divorce issue were people who argued that marriage was a contract, and that parties to any contract had the right to dissolve it. They also maintained that divorce was not the root cause of family disintegration. Rather, they saw divorce as a symptom, not a disease; as a cough is to a cold. Divorce was little more than a sign of turmoil and transition in the American family. Divorce was after the fact; it was the final seal of a couple's need to separate rather than the reason for their decision. Consequently, divorce was a result rather than a cause of changes in the institution of the American family.

Supporters of divorce often hoped that ease of divorce would eventually lead to equality and reciprocity in marriage. A growing number believed that divorce was a citizen's right in a democratic society. If divorces were easy to obtain for many causes ranging from adultery to mental abuse, there would seldom be reason for a couple to choose annulment, divorce of bed and board, or migratory divorce as a solution to their problems.

Even as Americans debated divorce, it gradually spread and became easier to obtain. By the mid-twentieth century, pro-divorce advocates had the satisfaction of seeing some of their ideas incorporated into law and policy in the United States.[3] At the same time, the influence of the anti-divorce forces waned. Granted, pro-divorce spokespeople were articulate and their arguments were forward-looking and modern, but anti-divorce leaders were also persuasive and their arguments for stability and continuity in American society had much to recommend them. Why then did the pro-divorce faction pull ahead of the anti-divorce faction?

The pro-divorce faction had several powerful allies; factors that pushed the divorce rate upward and forced many Americans to accept the presence of divorce. Sociologists and historians have long maintained that these forces included industrialization, urbanization, increasing mobility of Americans, broad-minded attitudes in the American West, men's and women's entry into the paid labor force, women's changing roles, and the gradual broadening of divorce laws and judicial decisions regarding divorce. Recently, several historians enlarged the list by demonstrating that the changing nature of the patriarchal family, rising expectations of marriage, and inequalities in relationships between husbands and wives also created marital tensions and divorce was often the result.[4]

In addition, the rising divorce rate itself played a role. The sheer force of numbers of divorces convinced some people, and compelled others, to accept the institution's increasing ubiquity in American society. Although the divorce rate had been rising since the mid-1600s, it reached an unexpected high during the 1880s: one out of fourteen to sixteen marriages ended in divorce.[5]

One hundred years later, the divorce rate had climbed to a new zenith: during the 1980s approximately one of two marriages ended in divorce. Americans, a people who love weddings, romance, and living happily ever after, had generated the highest divorce rate in the world.[6]

The pro-divorce faction had yet another advantage on its side: the concept of divorce was harmonious with American social and political ideas. Divorcing couples split apart to resolve their disagreements, just as numerous Americans solved other types of conflicts by leaving them behind. Perhaps more than any other national group in history, Americans have long exhibited a willingness to break unsatisfactory bonds and seek potentially more satisfying ties despite the costs.

This tendency to deal with disharmony by taking one's leave appeared early in the nation's history. During the 1600s and 1700s, numerous colonists left their families and homelands behind in search of less restricted, and more fulfilling, lives. In 1776, colonists who believed that their relationship with England hampered this quest, severed their long-term bond with their mother country. In the Declaration of Independence, which was a divorce petition of great magnitude, discontented Americans listed their grievances against the government of England, proclaimed their right to free themselves from a difficult situation, and declared their intention to seek greater happiness than they experienced within the British empire.

After the Revolution, Americans continued to exchange the old for the new, notably by trekking first over the Appalachian Mountains and then crossing the Mississippi River to develop new regions of the American frontier. They also persisted in demonstrating their commitment to personal freedom, individual fulfillment, and the intangible goal of happiness.

During the formative years of the new nation, a growing number of wives and husbands sought divorces. Then, as now, divorce fit well with American democracy and individualism. Divorce allowed people to make choices and reorder their lives when they deemed it necessary. It also underwrote the pursuit of personal happiness as a desirable goal. Gradually, proponents of divorce began to maintain that divorce was a citizen's right in democratic America: a civil liberty rather than a social ill.

As the great American debate between the anti-divorce and pro-divorce factions ebbed and flowed, legislators adjusted, and usually expanded, divorce legislation. They sometimes created compromise legislation to please opposing factions, but other times they simply translated prevailing ideas about divorce into law.[7] Reforms were often hasty, ill-conceived, and adopted under pressure from whichever faction had momentary influence with a particular group of legislators. As a result, divorce laws and policies often negatively affected the very people they were supposed to help: divorce-seeking men, women, and their children.

As divorce became increasingly common in the United States during the late nineteenth and early twentieth century, many Americans realized that the divorce process was too often inadequate, inconsistent, and even harmful to divorcing spouses and their children.[8] In addition, crucial questions regarding alimony and child custody went largely unexamined and unanswered. Judges often had the power to make alimony and child custody decisions as they saw fit. People who saw flaws in this system were often stymied in their attempts to improve the situation: they were unable to garner sufficient support in a society sharply divided in its attitudes toward divorce.

In the meantime, divorce itself was becoming more complicated. Each year thousands, then millions, of Americans experienced major life changes as a result of divorce. Even people who were happy to leave their mates, profited financially by divorce, or remarried immediately had to undergo wrenching readjustments. Moreover, frequent coupling, splitting, and recoupling created disconcerting social changes that demanded attention and adjustment. One minor but widespread example: as family members parted and joined with others to form new units, the branches of family trees became so crowded with former and present spouses, in-laws, and children that they groaned and cracked under the strain. By the late twentieth century, such concepts as family "line," passing on a family name, and family reunions were undergoing modification, while such pressing concerns as grandparents' rights and children's relationships with step-parents increasingly gained attention.

Other questions remained largely unresolved, including equitable ways of dividing a couple's property, determining support for financially dependent wives after divorce, and making child custody awards. During the 1970s, no-fault divorce was hailed as a welcome revolution, but by the 1980s, it was apparent that no-fault provisions contributed to the poverty of growing numbers of women and children in the United States.[9]

Perhaps most distressing, and least understood, was the effect of divorce on children. As more couples divorced, more children were separated from one parent or the other. During the late twentieth century, social workers, psychologists, and other experts who studied the effect of divorce on children reached disparate conclusions. Some maintained that children fared better in divorce than in a conflict-ridden marriage, while others presented evidence that divorce exacted its own heavy tolls of children. The results of a long-range study published in 1989 indicated that the negative impact of divorce on children was far greater than anyone had previously suspected.[10]

Remarkably, the American family remained resilient in the face of the rising divorce rate and its accompanying stresses. Increasingly, it seemed unlikely that the prediction that divorce would destroy the family would come true. Although some Americans decided against marrying and bearing children, and others delayed these events, the vast majority of Americans continued to marry and have children.[11] Yet other Americans cast marriage and family in a wide variety of non-traditional forms, including blended, single-parent, complex, and gay and lesbian families.

In the early 1990s, it seems clear that the American family will survive the growth of divorce. But will American law and policy prevail over the problems associated with divorce? Because divorce is evidently here to stay, it needs to be reshaped, to be made into the most expedient, beneficial process possible for divorcing men, women, and their children. If divorce is indeed a result rather than a cause of deep-seated alterations in the family, then it must help rather than hinder those family members caught in its maelstrom.

Thus, the depiction of American divorce that follows has several objectives. It traces the development of divorce from early colonial times to 1990, drawing upon a wide range of divorce laws, statistics, divorce and family records, personal interviews, newspapers and magazines, reformers' treatises, self-help literature, essays and novels, legal and sociological studies, and historical accounts. It is not a compilation of divorce laws, a statistical documentation of the rising divorce rate, or an exploration and exposition of the causes of divorce. Rather, it is a historical overview that provides a context for divorce as Americans know and experience it today.

Along the way, it points out some links between national ideology and individual values, maintaining that Americans' tendency to solve problems by splitting apart and beginning anew created a hospitable environment for divorce. Although it is unlikely that democratic ideals were directly related to divorce, they were compatible with oft-stated rationales for divorce.[12]

This account of divorce also reveals important aspects of the long-term debate regarding divorce, especially the widespread fear of migratory divorce. It argues that although the divorce debate interfered with the development of consistent, helpful divorce in the past, it need not do so in the future. Divorce can be reshaped into a more helpful institution.

This book, then, is about more than the past. It describes the course of American divorce since the early 1600s, and it provides a historical basis for understanding and dealing with divorce more effectively in the future.

# 1

# "Reward—
# Runaway
# Wife"

☐ There is no doubt that marital conflict existed in colonial America. Ministers and magistrates regularly heard cases of marital upheaval, while newspaper notices featuring such eye-catching leaders as "Reward—Runaway Wife" disclosed the dark side of married life to anyone who cared to read them. In these notices, abandoned husbands repudiated their wives' debts, while both husbands and wives appealed to the public for help in locating mates who had "eloped" or "absconded."[1]

Puritan leaders were especially distressed by the existence of troubled marriages because they fervently hoped to establish social harmony in the new world. Many of them believed that embattled spouses should either be reconciled or released. Although they did their best to coerce couples into remaining wed, they also instituted divorce proceedings for extreme cases of marital disintegration.

By permitting divorce, Puritan leaders initiated a democratic innovation with far-reaching effects. They offered American colonists, whether rich or poor, upper- or lower- class, male or female, the possibility of remedying excessive marital incompatibility and changing their lives. In other New England colonies and in the middle colonies, leaders experimented sporadically with divorce, but in the southern colonies, legislators and policy-makers eschewed it altogether. Yet, despite the reticence, or outright opposition, of many colonists, divorce gained a firm foothold in several parts of America by the outbreak of the Revolution in 1776.

Many Puritan marriages are revealed as less than ideal through New England church records. Marriages were often contracted under duress; an untold number, for example, resulted from a couple committing the sin of fornication which led to pregnancy followed by a forced marriage. Plymouth church records disclose that ministers regu-

9

larly publicly chastised couples whose first child was born as few as six weeks after their marriages. In the New Haven colony, fornication could lead to marriage even if no pregnancy occurred. A 1643 statute declared that fornicating couples be "punished" by a fine, corporal punishment, or "by enjoyning marriage." In that year, Margerett Bedforde was convicted of fornication and sentenced to be "severely whipped" and "marryed to Nicholas Gennings with whome she hath beene naught"—hardly an auspicious beginning for married life.[2]

Impotency and bigamy were other destructive conditions that troubled some Puritan marriages from the outset. For instance, in 1756, Plymouth church elders appointed a committee to "inquire" into Lydia Barnes's marriage to Jonathan Samson, formerly her niece's husband. Although church authorities eventually dropped the inquiry because its "consequence might otherwise be greatly to her Prejudice & Hurt," they believed that Lydia's marriage was bigamous.[3]

Church records show that numerous difficulties also erupted after marriage. Lydia Phillips was only one Plymouth woman who was censured by church elders because she had borne a child although her husband had been "a long time absent, if not dead." Even when spouses were present, some mates strayed. During his wife's pregnancy, Thomas Robinson of New Haven attempted to seduce several women. When he was brought to trial, one of his victims testified that he had "put down his breeches, put his hands under her coats, & gott them upp, thrust her to the wall or pale & indeavored with his boddy to committ adulterie with her, but she resisted & hindered him."[4]

Typically, Puritan ministers rebuked erring mates and coerced alienated couples to reconcile, sometimes against their wills. But some Puritans feared that forcing all estranged couples to remain harnessed by law would eventually undermine the social harmony they were trying to achieve. Adulterous relationships, non-functional families, physical and verbal abuse, and abandonment would surely result. They especially feared the destructive effects of adultery, which minister John Robinson called a "foul and filthy sin" and a "disease of marriage."[5]

Thus, such early and powerful proponents of divorce as ministers Cotton Mather and John Robinson and the ministers of the influential Cambridge Association, which argued for the liberalization of divorce during the 1690s, were reacting to reality. Disintegrating marriages created social discord and demanded regulation. Given their desire for order and harmony, colonial ministers and other leaders must have felt compelled to confront the problem of unworkable marriages; to control what Comfort Wilkins of Boston described as the "Tears, and Jars, and Discontents, and Jealousies" that afflicted numerous marriages.[6]

Pro-divorce Puritans agreed that embattled spouses should be encouraged to remain together, but they believed that divorce was a preferred solution in irremediable instances of marital breakdown. Divorce could dissolve highly dysfunctional marriages while controlling the terms and processes of parting. And it created the opportunity for divorced spouses to form more stable and orderly marriages in the future.[7]

Puritan leaders hoped that divorce would ultimately preserve the institution of the family. In the Puritan view, the family was the basic unit of society, crucial to social order and continuity. Marriages had to be sound if the community was to be sound. Marriage was also good for the individual, for it protected him or her from vice, especially by providing a setting for sexual activity. Consequently, although most Puritans preferred to keep marriages intact, some believed that irremediable cases should be eliminated by divorce. Rather than harm the family, they hoped to promote it with the safety-valve of divorce.[8]

Thus, they adapted to their needs portions of Continental laws regarding absolute divorce (*a vinculo a matrimonii*) and English canon law regarding bed and board divorce (*a mensa et thoro*). Because Puritans were far from united on the issue, it is likely that some people's opposition to the granting of absolute divorce impeded the process of developing divorce statutes and practices, while other people's support for divorce and the continuing necessity to deal with disintegrating marriages encouraged the process. As a consequence, early American divorce law grew sporadically and inconsistently.

Most other colonists did not follow the Puritans' lead. Rather, the majority of American colonists preferred to adhere to English practices of the 1600s that prohibited divorce and allowed ecclesiastical courts to grant only divorces of bed and board. This view especially held sway in the southern colonies, where leaders, who were predominantly Anglican and shared the church's belief in keeping marriages intact, refused to permit divorce.[9]

The chasm between pro-divorce and anti-divorce colonists planted the seeds of a debate concerning the pros and cons of divorce that swelled during the nineteenth century and raged well into the twentieth. In addition, it led to the establishment of widely divergent laws and procedures in each colony.

## New England as the Seedbed of Divorce

Separatists in Plymouth were the first Puritans, and the first colonists, to permit divorce in America. Before migrating to the colonies in 1620, many Separatists embraced Martin Luther's and John Calvin's belief that marriage and divorce were civil concerns. They also absorbed liberal Dutch views favoring divorce during their sojourn in Holland between 1608 and 1620.[10]

In 1620, Plymouth officials declared marriage to be a civil rather than an ecclesiastical matter. Plymouth's second governor, William Bradford, explained that marriage was to be "performed by the magistrate, as being a civil thing."[11] Although the declaration that marriage was a civil contract did not guarantee the establishment of civil divorce, it made it possible. In so doing, Separatist leaders defied English colonial policy which stipulated that English laws—in this case anti-divorce laws—be established in parts of the empire that lacked pre-existing legal codes.

Surviving records indicate that Plymouth Separatists granted at least nine divorces in the colony's seventy-two-year history: four on the grounds of adultery, three on desertion, one on bigamy, and one on incest.[12] Separate standards for men and women were non-existent; either could divorce the other for adultery. In addition, Plymouth courts also granted a number of divorces of bed and board, especially for a husband's abuse of his wife.

Yet, it was the Massachusetts Bay Puritans who granted the first divorce in the American colonies. When they began to settle north of Plymouth in 1630, they lodged jurisdiction for divorce in the Court of Assistants, composed of the governor, deputy-governor, and annually elected assistants or magistrates who sat as a court biannually, served as the upper house of the legislature, and acted as the governor's council.

According to fragmentary records, the first divorce in Massachusetts Bay, and in colonial America, was granted nearly a decade late. In December 1639, James Luxford's wife asked for a divorce because Luxford already had a wife. An unidentified magistrate granted a divorce and took this hapless woman and her children under the court's protective wing by seizing Luxford's property and transferring it to her. Next, the court turned its wrath on the deceitful Luxford. Not content with levying a fine of £100 on the bigamist, it sentenced him to "be set in the stocks an hour upon the market day after the lecture," and to be banished to England "by the first opportunity."[13]

Nearly five years passed before another divorce occurred. In March 1643, Anne Clarke acted on the precedent set by the former Mrs. Luxford and laid her marital complaints before the Court of Assistants. After Clarke explained that her husband, Dennis, had deserted her and was living with another woman, the court granted Clarke a divorce.[14]

Puritan officials eventually codified existing divorce procedures in a 1660 statute which explicitly vested the authority to hear divorce actions in the Court of Assistants. Subsequent court records show that the Court of Assistants accepted female adultery, male cruelty, bigamy, desertion, failure to provide, and impotence as grounds for divorce.[15]

Adultery was an especially thorny issue for the Bay colonists. Because their society was agrarian and property-based, female adultery was potentially a great danger: a pregnancy resulting from adultery might transfer a husband's property to another man's child. As a result, a woman who committed adultery was punished by a fine, whipping, being forced to wear the letter A for adulteress, being branded on the hand or forehead with the letter A, or being put to death. A man who committed adultery was punished only if his liaison involved a married or a betrothed woman. After consultation with church elders, officials agreed that female adultery was also a ground for divorce.[16]

Because the court was far less willing to grant divorces for male adultery, the few wives who charged their husbands with adultery usually bolstered it with other accusations, especially desertion and cruelty. The fate of one of the few women to charge adultery alone did not encourage other women to follow her path. In 1655, the Court of Assistants gave Joan Halsall a divorce after she charged her husband, George, with "abusing himself with Hester Lug" and branded him "an uncleane yoake-fellow." George Halsall appealed the divorce in 1659 to the General Court, consisting of the governor, deputy-governor, assistants, and annually elected town deputies that served as the Supreme Court and the lower house of the legislature. George Halsall requested a reversal of the divorce judgment and the return of his wife. The General Court voided the divorce and declared, presumably much to Joan's chagrin, that George could "have and enjoy the said Joan Halsall, his wife, again."[17]

In addition to acting as an appeal body in such cases as that of the Halsalls, the General Court also granted divorces in its own right. When it responded to William Palmer's 1650 petition accusing his wife, Eleanor, of deserting him and illegally remarrying, the General Court issued the first legislative divorce in Massachusetts Bay.[18] This action may have reflected English policy, established in the Act of 1534, that only the legislature—Parliament—could grant an absolute divorce. The coexistence of judicial and legislative divorces in colonies permitting divorce became common and added another complicating feature to American divorce procedures.

The year 1691 was a turning point for the fledgling institution of American divorce. When English authorites united Plymouth and Massachusetts Bay into the single colony of Massachusetts, these two jurisdictions had to combine and clarify their divorce statutes and procedures. In 1692, Massachusetts authorities adopted legislation reaffirming the principle of civil marriage and divorce. They also vested primary jurisdiction in divorce suits in the Court of Assistants.[19]

Massachusetts divorce records for the post-1692 period reveal several significant trends. The General Court continued to grant a small number of legislative divorces after 1692, institutionalizing a dual system of judicial and legislative divorce in the colony. In addition, desertion was seldom used as a ground for divorce because a 1695 law, "Act Against Adultery and Polygamy," allowed people whose mates disappeared at sea or were gone seven years without communication to petition the governor and Council for the right to remarry. The act drew upon the Cambridge Association's argument that "as for married persons long absent from each other, and not heard of by each other, the government may state what *length of time* in this case, may give such a presumption of *death* in the person abroad, as may reckon a second marriage free from scandal." In 1698, the requirement was reduced to three years in cases of absence at sea.[20]

Because colonial authorities generally refused to recognize the validity of slave marriages, few African-American colonists sought either absolute divorces or divorces of bed and board from Massachusetts courts. One of the few examples of an African-American divorce occurred late in the colonial period. In 1745, a Massachusetts slave man obtained a divorce on the ground of his wife's adultery. In 1768, an African-American divorce of bed and board occurred, when free black Lucy Purnan sued on the ground of extreme cruelty. In his most revolting exploit, her husband forcibly kidnapped her and sold her as a slave. She received the requested decree.[21]

Examination of Massachusetts Bay divorce records from before 1691 with those after that date reveals yet other characteristics. One of these was the adversarial nature of divorce suits. In colonial divorces, as in English divorce of bed and board hearings, one spouse was required to prove that the other had violated marriage vows. The requirement to assign blame in divorce suits replicated the procedure in other legal suits: one party charged the other with defaulting on an agreement. Consequently, a husband and wife became litigants, one suing the other for committing a breach of their marital contract. Courtroom battles frequently erupted from the requirement to assign fault and sometimes escalated into acrimonious appeal hearings disputing a lower court's judgment. Bitterness and enmity grew as friends, children, and other relatives entered the fray as witnesses, usually called deponents.

Once fault was determined, judges ordered harsh punishments for the "guilty" party who had caused a marriage to end and a family to separate. Undoubtedly, such punishment was also intended to deter other husbands and wives from committing similar acts. Typically, penalties were fines, whippings, and incarceration in the stocks. Divorce orders also usually prohibited the erring spouse from remarrying; this would keep a person who had been proven deficient in marital skills from creating another malfunctioning family unit. In some cases, the court banished or deported the sinner, presumably to free the community of a man or woman who had failed to demonstrate suitability as a member of the all-important unit of Puritan society, the family.[22]

The legal requirement to assign fault in divorce suits especially affected women, for the attribution of fault determined whether divorcing women could request financial support. Such support was usually called alimony, a term first used in England during the 1650s, meaning a financial allowance the husband paid his wife after their separation. If a wife was the plaintiff, she could ask the court to order alimony. Because she was free of fault, she deserved to have the financial support that her husband had pledged at the time of their marriage.

In Massachusetts, the first provision for alimony seems to have been a 1641 Massachusetts law stating that a divorced wife, if the "innocent" party, would retain her dower rights—a widow's share for life to one-third of her husband's estate. In 1679, Mary Lyndon fared well when a Massachusetts court granted her a divorce and ordered her former husband to give her two-thirds of his estate and other resources for her support. Yet, during the 1700s, only two divorce actions ordered alimony for wronged wives, perhaps because judges expected, and even intended to encourage, these women to remarry.[23]

If a wife was the defendant, she lost any chance of receiving alimony because legally her misbehavior had caused the marriage to end. She thus forfeited rights to her husband's financial support. Given this policy, it was advantageous for a woman to be a plaintiff rather than a defendant, or to protest her husband's charges so vociferously that the case was thrown out of court. Connecting the issue of alimony and the assignment of fault intensified the trauma involved in divorce proceedings, an unfortunate alliance that remained in force in America until the concept of no-fault divorce emerged in 1969.

Divorce was far from a nirvana for those Massachusetts women who chose to use it to end their marriages. In addition to loss of financial support, a woman might also lose custody of children and the respect of her family and community. As a divorced woman, she would have little freedom, for she would be expected to live with her parents or the family of a male relative rather than maintain a home of her own.

Still, divorce offered Massachusetts women an alternative that was out of the reach of dissatisfied English wives at the time. In 1700, an English feminist, Mary Astell, bemoaned women's lack of escape from unhappy marriages; they could only "bear their misery" and submit to the authority of their husbands. It was 1801 when the first English woman petitioned the House of Lords for a divorce. Her plea met with success, but failed to initiate a noticeable movement by other women.[24]

Despite the drawbacks of divorce for women, Massachusetts records indicate that a slowly growing number of women were willing to try to escape discordant marriages. During the 1600s, more men than women filed for divorce. But during the 1700s, women petitioners outnumbered men, especially after 1773, when Massachusetts officials offered women greater accessibility to divorce by showing a willingness to accept male adultery as a ground. According to a recent study, between 1765 and 1774, twenty-nine men and eighteen women requested divorces, but between 1775 and 1786, fifty-three women and thirty-three men did so.[25]

The rise in numbers of divorcing women may have partially resulted from increasing expectations of marriage. Divorce petitions and other records indicate that family life and people's attitudes toward the family were undergoing change as early as the mid- to late 1700s. The wording of women's divorce petitions during this period indicates that their expectations, which usually focused upon the basics of financial support and fidelity, began to include such intangibles as "affection," "nuptial happiness," and "love."[26]

The Lufkin divorce trial of 1760 is one example of a woman demanding love in her marriage. After Tabitha Lufkin reportedly seduced William Haskell, Stephen Lufkin sought a divorce on the ground of adultery. Haskell testified that he had indeed had sexual relations with Tabitha. Haskell explained that Tabitha had assured him that if he declined her offer, she would find "Some other Man For She Did not Love her Husband Lufkin."[27] Tabitha's aggressive pursuit of Haskell was the central issue for the court, which granted Stephen the divorce he requested, but Tabitha's use of the word love was also important. Most Puritans believed that love developed after marriage rather than as a prerequisite to it. If love failed to grow, couples were expected to stay together, bonded by their cooperation as economic partners and parents. Yet Tabitha justified her adultery because of the absence of her love for her husband.

During the 1600s and 1700s, Massachusetts courts granted a growing number of divorces. Between 1639 and 1692, thirty-one of the forty petitions filed for divorce or annulment appear to have been granted. Eighty-two of the 143 divorce petitions filed between 1692 and 1774 met with success in the form of a divorce, annulment, or separation.[28] But these figures represent only a portion of the divorces granted, for divorce records are incomplete, especially those from the 1600s.

Most contemporary observers regard the number of divorces granted by colonial Massachusetts courts as low, but this judgment may reflect a modern bias. Although slightly under one hundred divorces may seem small from a modern perspective, in all likelihood they constituted a considerable number for a society that was dedicated to keeping families intact except in extreme cases, and was in a very early stage of experimentation with divorce.

Also, from a statistical perspective, Massachusetts Bay's divorce rate took a giant step when it went from no divorce to the first divorce. When the Luxford divorce was granted in 1639, it represented a one hundred percent increase over the previous year. Moreover, after 1730, the number of divorces granted grew at a faster rate than did the population. According to one recent study, during the first one hundred years of divorce, the rate had remained relatively stable, but by the 1730s, the rate began to rise.[29] It is likely that the colonial divorce rate was as shocking to most Americans then as contemporary divorce rates are now.

It is also important to consider the composition of the Massachusetts population when determining whether numbers of divorces were low or high. One estimate of the colony's population put it at 8,592 inhabitants in 1639—the year of the first divorce—and 360,000 in 1774. The number of divorces granted appears small against these figures. But these population figures included African Americans, mulattoes, and American Indians who only in rare cases divorced in white courts. The figures also included male children under sixteen and female children under fourteen who by law were too young to marry and divorce. Subtracting these groups from the total population leaves a much smaller pool of white adults who could marry and thus divorce. Against a reduced population figure, the number of divorces appears larger.

Because there was no Massachusetts census in those years, it is impossible to know how many Massachusetts citizens were married—and thus eligible to divorce. A 1765 local census suggests that the number of families was fairly low. Boston, for example, encompassed 15,520 inhabitants, but only 2,069 families. Hingham had 2,506 inhabitants and 426 families. These figures suggest that colonial Massachusetts contained a relatively small pool of marriages to have produced as many divorce petitions as it did.[30]

Rather than debating whether the number of Massachusetts divorces was low or high, it is perhaps more important to consider the trend that those divorces established. The number of divorces rose intermittently before 1765 and markedly after that year, thus establishing an upward pattern that continued during the 1800s. When the U.S. Census Bureau collected marriage and divorce statistics in 1888, it estimated that one out of fourteen to sixteen marriages ended in divorce.[31] The American divorce rate did not jump to this level overnight; rather, it was a legacy of a pattern established during the colonial period.

We must also remember that divorce decrees do not measure the extent of marital breakdown. Numbers of unsuccessful petitions always outdistanced numbers of divorce decrees, often by a two to one ratio. Scattered evidence suggests that many more people may have thought about divorce, but ended their marriages through desertion and separation instead. Although desertion appeared to have been widespread, it escaped inclusion in formal records. Separations also remained uncounted although they occasionally appeared in court records. For instance, in 1666 and again in 1676, Mary Drury of Suffolk County was brought before the court for refusing to live with her husband, for being a "runaway wife." On both occasions, Mary was fined and ordered to return to her husband. The court denied her subsequent request for a divorce and ordered her to reconcile with her husband. Perhaps Mary remained in her marriage, but it seems likely that she separated from her husband on subsequent occasions.[32]

Unquestionably, Massachusetts statutes and policy significantly affected the development of American divorce, but leaders in other New England colonies also experimented with divorce and contributed to its evolution. Connecticut leaders especially held views similar to their Massachusetts neighbors'. Because they too were passionately committed to maintaining order in society and in families, they frequently expressed dismay regarding the marital discord that threatened the harmony of their colony. Four years after Connecticut's settlement in 1636, the colony's lawmakers voiced their impatience with mismatched couples: "Many persons entangle themselves by rash and inconsiderate contracts for the future joining in Marriage Covenant, to the great trouble and grief of themselves and their friends."[33] To the concerned legislators, divorce seemed a reasonable way of eliminating severe cases of family turmoil and safeguarding the family.

During Connecticut's early years, the General Court, a body wielding both legislative and judicial power, granted divorces. The first was in 1655 to Goody Beckwith on the basis of her husband's "departure and discontinuance."

The separate government of New Haven enacted slightly different divorce legislation. New Haven statutes gave the Court of Magistrates, a biannual court composed of magistrates from all jurisdictions, the right to grant divorces. The statutes established the first formal list of grounds for divorce in colonial America: adultery, desertion, and male impotence, or, in the tactful words of the legislators, a husband's failure to perform his "conjugall duty" to his wife.[34]

A 1657 case demonstrates that people did divorce on the ground of impotency. After John Vffoote's wife divorced him for "insufficiency," he was charged with committing fornication with Martha Netleton, his father's servant. In his defense, he claimed that his wife had failed to act "toward him as she ought," but that Martha had restored his faith in himself. After fining John and Martha for fornication, the court granted their request to marry.[35]

This case also shows that love was already an issue for some people. In a subsequent hearing investigating Hannah Beard's cool treatment of her former husband, John Vffoote, testimony disclosed that Hannah thought John a "foole," that she felt no "wife like affection" for him, and that she believed it was "pittious" that she had to live with a man she had never loved. Although the court reproved Hannah for her irresponsible behavior, it levied only a small fine on her.[36]

Connecticut and New Haven had to begin considering ways to reconcile their divergent divorce statutes when they united into a single colony in 1665. Two years later, members of the Connecticut General Assembly adopted divorce provisions similar to New Haven's. The Court of Assistants, later known as the Superior Court, received the power to grant divorces on the grounds of adultery, fraudulent contract, willful desertion of three years, and absence of seven years' duration. The General Assembly also continued to grant divorces and served as the court of appeals in contested divorce decrees.[37]

Unlike Massachusetts courts, Connecticut refrained from giving divorces of bed and board.[38] In all likelihood, judges and others opposed such decrees because they inadvertently punished the innocent party by prohibiting remarriage. Divorces of bed and board also forced a man to support a woman he no longer lived with, or sometimes harnessed a woman to a man who refused to pay court-ordered alimony, while restricting her from either remarrying or becoming self-supporting.

Many Connecticut divorce suits reached quick and easy resolution, for the Superior Court frequently granted a divorce without a full-blown hearing. Cases based upon desertion, the most frequently used charge in the colony, could be easily dispatched because the deserting spouse, often said to be residing in "parts of the world unknown," was absent from the divorce hearing. In 1723, Sarah Whelsher swore that she had heard nothing from her husband for three years. She added that her children were destitute as a result of his abandonment and that she faced poverty and disgrace. The judge promptly granted Whelsher a divorce.[39]

Speedy resolution was called for in desertion cases so that an abandoned spouse could remarry and form a new family. Also, abandoned wives needed to be extricated from their married status as *feme covert,* which prohibited them from owning property, establishing businesses, signing contracts, or in other ways managing personal affairs or supporting themselves. Once divorced, women returned to the status of *feme sole,* which allowed them to conduct personal business and seek a livelihood.

Other more complicated suits, however, led to court trials that were sometimes long and nasty. One acrimonious trial began in 1716 when John Merriman petitioned the Superior Court for a divorce on the ground of his wife's desertion. Hannah Merriman responded with the caustic assertion that John had "broken most if not all the essential articles in his covenant of marriage" with her. During the ensuing trial, John resorted to the common practice of asking ministers to speak against a wife's desertion of her husband. Hannah retaliated by presenting eyewitnesses to John's cruel treatment of her. She vowed that she would rather "live in the jail all her days" than live with him. After months of delays and conflicting testimony, the court ruled that when fault existed on both sides and one party was unable to prove his or her own innocence and the other spouse's guilt, no divorce could be granted. This controversial principle would be adopted by other courts during succeeding years, but John was unwilling to accept it. He appealed the judgment and in 1718 finally received the coveted decree.[40]

Some years later, a similar, but far more notorious divorce trial, reached the Connecticut Superior Court. The infamous Thrall case was initiated by William Thrall in 1735 on the ground of desertion. His wife, Hannah, had left him three years earlier because of his abusive behavior. Determined to divorce Hannah, William attempted to strengthen his desertion charge by adding an allegation of adultery. Although he was unable to prove this accusation to the court's satisfaction, his attempt provided a bit of courtroom drama. Hannah aggressively countered William's charges with her own complaints, undoubtedly mindful that she would lose her dower rights to William's substantial property if the court judged her the guilty party.

Complex and hotly contested, this tempestuous divorce trial dragged on for years, finally ending when the court reached two conclusions. One was the general principle that a husband who drives his wife away is himself the deserter. The other was the specific principle that William Thrall did not fit the New England ideal of a stable, patient husband and would not get a divorce. The traumatic

Thrall marriage ended in 1739 when William died, leaving Hannah financially independent at age thirty-six.[41]

In colonial Connecticut, divorce suits involving adultery and bigamy virtually always went to trial. Men more frequently employed the complaint of adultery against their wives than did women against husbands, presumably because courts found female adultery a more heinous crime than male adultery. Husbands seldom hesitated to introduce revealing evidence, including eyewitness accounts. In 1775, Reuben Benham submitted depositions from his son and the family's maid substantiating his wife's adultery with several men. Other male petitioners testified that they had returned from extended sea voyages or other absences to find newborn infants in their wives' arms.[42]

Women, however, alleged bigamy on the part of husbands far more than men did against wives. Also, they often added cruelty as a secondary, substantiating charge, thus laying the basis for later generations' acceptance of cruelty as a ground for divorce.[43] Charges of cruelty, which were perhaps introduced to elicit judges' sympathy, almost always described physical rather than mental abuse. Mehitable Griswold's 1746 petition was typical. She accused her husband of treating her "unkindly," neglecting the "Marriage Bed," and threatening her life. Two witnesses added that they had seen her "with both her hands tyd together and her arms bludy." In 1757, Sarah Johnson testified that her husband tried to control her by "beating, striking, kicking and stamping upon her, pulling her hair and buffeting and spitting in her Face in a most angry, malicious manner." Beginning in the mid-1700s, however, a few women accused their mates of verbal cruelty. In 1755, Anna Castle related a disquieting tale, alleging that her husband had repeatedly sworn at her, called her vile names, and even taught his children from a previous marriage to call her "whore" and other epithets.[44]

The few male petitioners who added cruelty charges to their accusations against their wives focused more on verbal than physical abuse. Elijah Metcalf complained in 1771 that his wife, Mary, bombarded him with vituperative remarks, including an assertion that she "did neither love nor fear him and did not care if the Devil had him." But, also in 1771, an instance of physical abuse came before the court when Jesse Higgins declared that his wife, Alice, struck him repeatedly.[45]

During the 1700s, Connecticut courts granted more divorces than Massachusetts courts although Connecticut's population was approximately 40 percent of that of Massachusetts early in the century and 75 percent of it in 1770. When Bostonian Sarah Knight traveled to New Haven in 1704, she soon became aware of Connecticut's lenient stance on divorce. She reportedly remarked that divorces "are too much in Vogue among the English in this Indulgent Colony as their Records plentifully prove."[46]

African Americans only infrequently figured in Connecticut divorce proceedings, although one estimate put the black population of Connecticut as high as 4,590 in 1762 and 6,464 in 1774. A rare example of an African American suing for divorce occurred in 1771 when a slave named James took his wife, Grace, a mulatto slave, to court; the outcome is unknown.[47]

Another New England colony that permitted divorce was Rhode Island, settled by dissenters from Massachusetts in 1636. Rhode Island authorities received the colony's first divorce petition in 1644. Six years later, legislators passed the colony's first divorce statute which allowed the General Assembly to grant divorces on the single ground of adultery. Subsequent cases indicate that in fact divorces were granted on the grounds of desertion and impotence as well. Presumably because it believed that divorce cases were more appropriately judicial matters, in 1655, the General Assembly shifted the right to grant divorces to local courts, but heard divorce pleas on grounds other than adultery. In 1667, legislators moved jurisdiction for divorce to the Court of Trials, composed of the governor and council. In 1747, legislators lodged jurisdiction in county courts.[48]

Rhode Island divorce records show that the majority of petitioners cited desertion as their complaint. Men, who were accused of desertion more often than women, usually fled to other towns, other colonies, or to "parts unknown." Female deserters chose similar havens, but women sought refuge with their families far more often than men did. Daniel Barnes indignantly protested that when his wife returned to her family in 1769, she was escorted out of his house by her father and brothers who took a generous share of furniture and household goods with them.[49]

Rhode Island divorce documents also offer insight into the intriguing matter of how plaintiffs proved to the court's satisfaction that their spouses had committed the seemingly private act of adultery. The wronged party usually relied upon friends, family members, and neighbors to act as witnesses to extra-marital affairs. Some witnesses testified that they had eavesdropped on incriminating conversations or heard a husband or wife brag about sexual exploits, but others gave eyewitness accounts. Such testimony was possible because colonists lived in close proximity to family members, friends, and neighbors. In addition, because they believed that they were their brothers' and sisters' keepers, colonists seldom hesitated to overhear or observe untoward behavior.

Testimony in adultery cases was often specific and highly incriminating. One typical adultery trial revolved around a young woman's description of an event that had occurred when she shared a bed with Sarah Staples. She testified that Ephraim Whipple "climbed into the other side" and "had Carnel Knowledge of the body of ye said Sarah." The deponent's fourteen-year-old brother added that he had seen Sarah and Ephraim "both in motion on the ground" on another occasion. A less representative adultery trial focused on a man's graphic portrayal of the night he and five other men and women spent "promiscuously in the same bed" during the absence of the husband of one of the women.[50]

Rhode Island documents also reflect the financial problems that bedeviled many divorcing wives. Women who accused their mates of desertion, adultery, or other misconduct frequently added a description of their own destitute circumstances. Women related their pitiful stories of relying on friends and family for bare subsistence, or of turning to poorhouses and other charitable institutions. Such pleas squarely indicted men for failing in one of their primary marital duties—financially supporting their wives and children.[51]

Information regarding divorce in the last New England colony, New Hampshire, is sparse. It is likely that New Hampshire's divorce practices were similar to those of Massachusetts, for New Hampshire was part of the Bay colony until 1680. The first divorce case heard in New Hampshire seems to have been a 1681 petition from Sarah Pearce that described her husband as an abusive man who threatened to kill her "by poison, or knocking of the head" and maintained that he eventually deserted her. Apparently, after New Hampshire separated from Massachusetts in 1680, the governor and council began granting divorces. The only formally designated ground was desertion. Beyond this meager information, the divorce history of colonial New Hampshire remains unknown.[52]

## Hesitancy among the Middle Colonies

Divorce was far more scattered in the middle colonies than in New England. Well-defined statutes were largely non-existent. In addition, divorce records are fragmentary and offer an incomplete picture of divorce in this region.

In New Netherland, which was founded in 1624 by a group of Dutch settlers under the auspices of the Dutch East India Trading Company, leaders supported Dutch ideas in favor of divorce. Piecemeal records indicate that Dutch authorities granted at least three divorces under their tenure: one in 1655 to John Hickes because his wife left him and "married" another man; one in 1657 to Joris Baldingh because of his wife's "dishonorable comportment" with another man; and one in 1664 to Annecke Adriaens because her husband took a second wife. Adriaens's twice-married spouse received harsh punishment for his misconduct: he was flogged, "branded with two marks on his back," and permanently barred from the colony.[53]

The year 1664 was not only traumatic for the husband of Adriaens, but for Dutch settlers in New Netherland as well. In that year, King Charles II of England gave the Dutch-occupied territory to his brother, James, the Duke of York. James sent troops to quell the feeble opposition of Dutch leaders and appointed his own governor of the colony, which he renamed New York. Although the Duke's laws, which James promulgated in 1665, said nothing explicit regarding divorce, they included a provision that "it shall not be punishable to remarry" in the event of a spouse's bigamy, death away from the colony, or unexplained absence of five years. A 1665 addition to the Duke's laws provided limited redress in adultery cases: "all proceedings shall be according to the laws of England," meaning that divorces of bed and board could be granted in the event of adultery.[54]

Despite these limited provisions, New York colonial records disclose that divorces were granted on grounds ranging from adultery to desertion. Several governors evidently believed that they had the power to grant divorces. One of these was Governor Francis Lovelace, who released Thomas Pettit from his marriage to Sarah in 1672. Sarah had allegedly defiled "the marriage bed" and committed adultery "with several persons." A few years later, Lovelace's successor granted a divorce to Catherine Lane after her husband raped and committed incest with their daughter. He ordered Catherine's husband to give her a portion of the family property to support herself and her children. Sometime around 1675, New York governors apparently ceased to give divorces; none appears to have been granted after that year.[55]

In the neighboring colony of New Jersey, which separated from New Netherland in 1664, the colony's governor granted at least three divorces. The grounds are unclear in two of the cases. In the first divorce, which occurred in 1669, the colony's governor declared Hopewell Hull and his wife divorced on the vague ground that he wed her "unlawfully." In the other unclear divorce, heard in 1673, Thomas Davies and Margaret Blew mutually requested that their marriage should be dissolved on the ground of "divers causes."[56]

An additional divorce was granted in 1692 by the Middlesex County Court of Sessions. This New Jersey court granted a divorce to Rebeckah Adams Seaton from her husband, James, on the grounds of desertion and living with another woman even though no formal divorce law existed in New Jersey at the time.[57]

Attitudes and actions regarding divorce in New Jersey after it became a royal colony in 1702 are even more obscure. Scattered evidence indicates that the governor, and perhaps the colony's legislature, continued to grant infrequent divorces, but New Jersey records are too incomplete to warrant any firm conclusions.

The government of Pennsylvania also enacted divorce statutes. Shortly after William Penn received a charter from King Charles II in 1681 for a Quaker colony called Pennsylvania, he formulated statutes that prescribed a "Bill of Divorcement" as part of the punishment to be meted out to adulterers. A subsequent act of 1700 added sodomy, bestiality, and bigamy to the list of crimes punishable by divorce. When the English Parliament disallowed this punitive use of divorce, Pennsylvania authorities devised other statutes in 1705 that allowed divorce in cases of consanguinity and permitted divorces of bed and board in cases of adultery, bigamy, and sodomy. These statutes also prescribed physical punishments for guilty spouses, including whipping, fines, branding, and imprisonment. Because Pennsylvania legislators neglected to establish clear procedures for implementing these laws, only a few absolute divorces and divorces of bed and board were granted before 1776.[58]

The divorce history of the last middle colony, Delaware, is vague. At least one person requested a divorce while Delaware was part of Pennsylvania. Although there seemed to be little question that the husband was an adulterer who frequently gave his wife a "severe drubbing" and turned her out of the house "as a dog," a final disposition to the case does not appear in the Pennsylvania annals. When Delaware separated from Pennsylvania in 1701, lawmakers devised a code of laws that apparently neglected to provide for divorce. Rather than being cause for divorce, adultery was to be punished by fines and whipping.[59]

## Southern Conservatism

The story of divorce in colonial America becomes somewhat simpler when the focus turns to the southern colonies. Although southern leaders, like those in New England, decreed that marriage was a civil matter, they refused to countenance divorce.

When the first Virginians arrived at Jamestown in 1607, they adhered to Anglican beliefs more closely than did the Puritans and thus transplanted English policy prohibiting divorce. They agreed, however, that divorces of bed and board on the basis of adultery, desertion, and extreme cruelty were legitimate. But Virginia lacked the bishops and ecclesiastical courts that granted divorces of bed and board in England. Thus, Virginia courts of chancery filled the void by hearing cases of marital upheaval. Following English ecclesiastical court practices, chancellors frequently ordered husbands to continue to support wives who were to live apart from them. Neither party was to remarry.

Women initiated virtually all separate maintenance requests in colonial Virginia. They usually wanted to live separately from their husbands and regain control of their dowries or receive a portion of their husbands' estates as support. By intervening on behalf of Virginia wives, courts of chancery brought desperately needed financial relief to wives imperiled by absent, cruel, adulterous, and bigamist husbands. But, separate maintenance decrees did little for husbands who could informally separate from wives without paying anything or paying only what they thought adequate, and thus, had nothing to gain by obtaining separation orders.[60]

Separate maintenance cases in Virginia typically involved charges of cruel behavior. In 1699, Mary Taylor sought separate maintenance because of her husband's abusiveness. She explained that she wanted a home where she would be "secure from danger." Taylor received furniture, clothing, and twelve hundred pounds of tobacco as alimony. The following year, Elizabeth Wildy, whose husband was a landholder and county official, maintained that he had beaten her, held her in the fire, and threatened to shoot her. The court ordered her husband to make an annual cash payment to her.[61]

In other southern colonies—Maryland, chartered in 1632 and dominated by Roman Catholics; the Carolinas, founded in 1663; and Georgia, founded in 1732—official views remained in harmony with English thinking regarding divorce. On several occasions, Maryland legislators discussed the possibility of granting divorces or allowing county courts to do so, but they rejected the idea.[62] Instead, as in Virginia, Maryland chancery courts heard cases of marital upheaval. Although these courts could not grant divorces of bed and board as did English ecclesiastical courts, they remained as faithful as possible to English practice by ordering separate maintenance for aggrieved wives.

South Carolina cases show that people from a variety of social classes requested separate maintenance orders. In 1723, a South Carolina court ordered Stephen Taveroon to give forty shillings a week to his wife whom he had physically abused. A decade later, chancellors required Charles Lowndes to "give Security for his good Behaviour" toward his wife, Ruth, and to furnish her with household goods, a sidesaddle, a slave woman and two slave children, and £50.[63] The discrepancy in these awards reflected differences in Taveroon's and Lowndes's resources as well as their wives' standards of living, factors often considered by courts when assigning separate maintenance.

Court-ordered alimony was not always paid, however. When Margaret Macnamara of Maryland obtained an alimony award as a result of her husband's cruelty, she discovered that he had no intention of giving her the clothing, personal items, and £15 a year that the court had ordered. Thomas Macnamara argued that the Court of Chancery lacked the jurisdiction to grant separate maintenance. But, after being imprisoned three times, Macnamara begrudgingly paid the stipulated amount to his wife.[64]

The lack of divorce in southern colonies does not mean that southern marriages were less troubled than other colonial marriages. On the contrary, travelers' accounts indicate that marital problems abounded among all races and classes of people. An agent for an English firm touring the North Carolina back-country in 1753 was appalled by the number of contentious couples that he encountered, and by a rural justice of the peace who openly admitted that two of his children had been fathered by another man.[65]

Southern court records of marital disputes and adultery and bastardy cases verify such observations. One example was North Carolinian Honora Donean, who had married in 1743. Shortly afterward, her husband, John, "turned her away" and refused to support her. Honora ingeniously established a slaughter pen where cattle could be killed for ten shillings near the New Bern marketplace. When John interfered with her business, she appealed to the North Carolina Executive Council for redress. Rather than urging the couple to reconcile, council members validated the Doneans' informal separation by issuing a restraining order against John. Honora was free to conduct her business and support herself.[66]

Southern newspapers also reveal the pervasiveness of marital difficulties. Advertisements for runaway wives, notices repudiating wives' debts, and announcements of informal separations disclose that many couples fought and parted despite the absence of formal divorce. In 1775, Coleman and Elizabeth Theeds announced in the *Virginia Gazette* that they had "parted by mutual Consent." Moreover, they had "given Bond each to the other . . . that they will not interfere with any Estate which shall hereafter accrue to either Party." Although such newspaper divorces were not legally binding, parties to them frequently considered them legal and even remarried.[67]

Marital discord afflicted lower-, middle-, and upper-class marriages throughout the colonial South. Propertied families, who were well aware that divorce was unavailable when trouble struck, attempted to shield brides-to-be. They negotiated marriage contracts to protect family property from incursions by husbands and to ensure wives some financial independence from difficult husbands. Contracts often limited income from a wife's property to her use and specified that the property the wife brought to the marriage would be hers after her husband's death. She would thus have financial resources if he had diminished or destroyed his own estate. Margaret Haynes of North Carolina was protected by a typical contract. It stipulated that her property would be for "her Separate use, benefit, and behoof" after marriage and was not liable for her husband's debts. When a woman had children from a former marriage, a contract might also designate a sum of money or piece of property for the children's support.[68]

Divorce may have been absent in the South, but formal and informal separations appear to have been widespread. In addition, scattered evidence suggests that divorce occurred among enslaved and free African Americans, and among American Indians. Although they were recorded in such scattered documents as church and plantation records, these divorces are an important part of southern divorce history that need to be retrieved from oblivion. Moreover, Louisiana was a French colony during the 1600s and 1700s, but its leaders were developing attitudes and policies that would become yet another motif in American divorce when Louisiana became part of the United States in 1803. During the years that southern officials in American colonies lacked the formal power to grant divorces of bed and board, the Superior Council of Louisiana regularly granted such orders. Surviving French colonial records are peppered with bed and board requests, rebuttals, and decrees.

Louisiana divorce of bed and board cases were virtually always based on the charge of cruelty. One such case occurred in 1728 when Louise Joussset La Louire of New Orleans petitioned the council for a divorce of bed and board. La Louire complained that her husband, surgeon Pierre de Manade, had "dissipated" her considerable dowry, regularly kicked and beaten her, and hit her with a hammer, stick, and other heavy objects. She had finally fled from him and took refuge in an Ursuline Convent. Although Manade refuted her charges and said that she could stay in the convent as long as she wished, the council ruled against him; it ordered him to give his wife property and goods and to return her dowry. Whether Louise received a divorce of bed and board, or ever left the safety of the convent, remains unknown, for the documents were charred by fire. In a similar struggle to retrieve her dowry and get some of her husband's property, another wife requested "hard" coin, but finally accepted one black slave, 2000 pounds of deer skins, and a small amount of paper money.[69]

An upper Louisiana case followed the same lines. Pelagre Carpentier Vallé of Ste. Genevieve obtained a divorce of bed and board from Charles Vallé because he allegedly drank, beat her, gambled, kept a mulatto mistress, and squandered their joint property. She received part of their property and other resources to support herself and their two children.[70]

Some French women's petitions asked only for the return of their dowries. Antoine Aufrere initiated a suit on behalf of his minor daughter, Françoise Pery, in 1743. Pery received a divorce of bed and board that ordered her husband to restore her dowry, including money, slaves, and cattle. Because her husband refused to comply, this case reappeared several times in the official record. Finally, the court ordered that the recalcitrant husband's goods be inventoried and sold at public auction. His wife finally received the share specified by the court.[71]

The unremitting theme of wife abuse in Louisiana divorce of bed and board pleas suggests that such documents usually focused on legally acceptable grounds. Thus, such petitions are untrustworthy accounts of the underside of marriage. Although abuse may have occurred, petitions failed to present a comprehensive description of a couple's marital conflicts.

A case illustrating this point involved a wife who testified that her husband was abusive. She pointed to several scratches and a facial bruise as evidence. Her husband had a different view of the situation. He responded that she had "no idea of the obedience a wife owes her husband," and that she had opposed a property transaction that he wanted to make on behalf of her son. He added that she had inflicted the scratches herself. She received the bruise when she put her ear to a door to overhear him. She was "struck by the latch" when he opened the door.[72]

While abuse was an issue in this case, this couple apparently also disagreed about a wife's power, especially concerning the disposition of property. In this case, the divorce of bed and board occurred because of abuse, combined with a struggle over gender roles, distribution of power, and property. Thus, such documents need to be read with care rather than as literal accounts of the causes of marital breakdown.

## Responses to Divorce

How did English officials react to growing numbers of divorces, separate maintenance orders, and separations among American colonists? Could English authorities accept divorce among distant subjects when English ecclesiastical courts gave only divorces of bed and board and Parliament granted a limited number of divorces? Would they allow Americans to flaunt colonial charters that clearly instructed colonists to follow English laws and practices?

Surprisingly, divorce in the American colonies seemingly went unremarked in England during the 1600s and the first half of the 1700s. Under imperial law, actions of colonial courts were usually free from review or approval by English authorities. Only colonial statutes had to go to the Board of Trade for ratification or rejection. Eventually, a number of divorces granted by colonial legislatures were submitted to the Board of Trade with other legislation.

During the late 1750s, the English Board of Trade reacted to several Massachusetts legislative divorces by calling them "extraordinary" and "liable to great objections." Board members disapproved of colonists' granting divorces on grounds other than the biblical one of adultery. They were upset that colonial legislators had seized the power to grant divorces. The Board of Trade recommended an investigation, but inexplicably none resulted. The divorces stood by default. As a result, Massachusetts authorities continued to ignore English rules and to grant divorces.[73]

Some years later, Pennsylvania authorities sent a batch of acts to England for approval, including the 1769 legislative divorce of Curtis and Anne Grubb. Members of the Board of Trade were again alarmed because colonial legislators had exercised a power that English law limited to Parliament. They sent the divorce bill to a prominent London attorney who responded that colonial legislatures should be allowed to grant divorces to colonists because Parliament was out of their reach. At his suggestion, the Grubb divorce bill was sent to several government officials for review, but when they failed to comment for reasons that are unclear in the record, the divorce went into effect.[74]

Three years later, the Pennsylvania legislature sent the Board of Trade the divorce act of George Keehmle, a Philadelphia barber who had accused his wife Elizabeth of adultery. On this occasion, members of the Board of Trade referred the bill to the King's Privy Council. In 1773, the King's councillors disallowed the Keehmle divorce. Presumably, this was distressing news to the Keehmles, who had been living as divorced people for almost a year. Shortly thereafter, the Royal Instructions to Governors of the Colonies ordered colonial governors to reject all divorce bills: "It is our express will and Pleasure that you do not upon any pretence whatsoever give your assent to any Bill or Bills. . . . for the divorce of persons joined together in Holy Marriage."[75]

This command probably added a small bit of fuel to the revolutionary fire that was beginning to blaze in the American colonies. When the Declaration of Independence presented the colonists' grievances to a stunned world in 1776, the first complaint protested that King George III had "refused his Assent to Laws, the most wholesome and necessary for the public good." Surely colonial authorities who supported divorce, and those colonists who wished to divorce, saw divorce as being for the "public good" and resented its disallowance by English authorities.

Between the founding of Jamestown in 1607 and the Declaration of Independence in 1776, a variety of social and economic factors had altered American marriage and encouraged some people to think about divorce. The growing mobility of colonial Americans and the gradual development of a westering fever put strains on marriages and families. The emergence of a market economy, technology, and new forms of production began to rob the family of its role as America's basic economic unit. By the mid-1700s, the economic partnership of husband and wife as producers of agricultural and other goods was beginning to show early signs of decline. During the 1760s and 1770s, these factors were exacerbated by resistance to English rule and the development of a rhetoric of individualism; a rhetoric that assured Americans that liberty, justice, and the pursuit of happiness were their right.

As the American Revolution neared, a number of colonists began deliberately to link revolutionary principles to divorce. One important example was Virginia attorney Thomas Jefferson, who related the concepts of independence and happiness with divorce some years before he presented a similar argument for terminating America's connection with England in the Declaration of Independence. Sometime in 1771 or 1772, Jefferson prepared notes for the divorce case of Dr. James Blair of Williamsburg, who wished to divorce his wife after nineteen months of turbulent marriage. In his notes, Jefferson tied revolutionary ideals to divorce in order to persuade the Virginia legislature to grant the colony's first absolute divorce. Although Jefferson never submitted the case to the Virginia General Assembly due to Blair's death, his plea employed the same principles that he, and many other Americans, would soon use for revolution.

Jefferson's notes suggest that he planned his strategy carefully, perhaps with hopes of presenting arguments that would appeal to legislators deep in the throes of resistance to England. His opening salvo concerned rationales for divorce. He maintained that it was cruel "to chain a man to misery till death." He then wrote that "liberty of divorce prevents and cures domestic quarrels," and "preserves liberty of affection." When Jefferson turned his attention to the nature of covenants, he asserted that contracts were unenforceable when they opposed the will of the involved parties. Moreover, unproductive contracts were "dissoluble." Ironically, although Jefferson was representing the husband in the Blair case, he viewed divorce as a boon for women. Under "Miscellaneous Observations," Jefferson commented that divorce "restores to women their natural right of equality" and that it was "cruel to confine Divorce or Repudiation to [a] husband who has so many ways of rendering his domestic affairs agreeable, by Command or desertion, whereas [a] wife [is] confined & subject."[76]

Jefferson's notes incorporated Puritan arguments for divorce, Lockean arguments for dissolvable contracts, and emerging American attitudes regarding women. He applied such terms as liberty, natural right, equality, and dissoluble contract to the case of an individual citizen caught in a distressing marriage, but he stopped short of advocating equality between men and women and the eradication of the patriarchal family.

Had he lived one hundred years earlier, Thomas Jefferson probably would have taken a much different stance. But, by the early 1770s, a century-long trend was in force: the gradual broadening of American divorce attitudes and practices in a number of colonies. In Massachusetts and Connecticut, for example, during the 1700s the number of divorce petitions had increased, the number of women filing for divorce had grown, and the success rate of women's petitions had improved noticeably. Given these changes, it is conceivable that Thomas Jefferson could prepare a divorce case in anti-divorce, Anglican Virginia. Once the American Revolution began, many divorce petitions had more to do with revolutionary ideology than with such war-related problems as adultery committed by separated spouses.[77] Arguments used in wartime petitions frequently mirrored the revolutionary fervor that enveloped the new nation: some petitioners maintained that one party could flee from another if their union lacked fulfillment and happiness. Divorce-seekers also employed such terms as "tyranny," "misrule," "injustice," and "happiness of the individual," while espousing revolution against unjust rule by a spouse. These arguments are even more striking when compared with English divorce petitions which concentrated upon matters of property rather than such slippery issues as liberty and happiness.[78]

The influence of revolutionary philosophy was more than a wartime phenomenon, however. Such people as Abigail Strong and John Christian Smith used revolutionary ideas to justify divorce after the war's end in 1783. In 1788, Strong declared that if "Kings may forfeit or discharge the allegiance of their Subjects," she also had the right to be freed from allegiance to her abusive husband. Smith drew upon the ideals stated in the Declaration of Independence in his 1791 petition. "Since Marriage was instituted for the purpose of promoting the happiness of individuals and the good of society and since the attainment of those objects depends entirely on the Domestic harmony of the parties connected and their living together in a perfect union of inclinations interests and affections," he reasoned, "when it becomes impossible for them to remain longer united . . . the good of Society no less than the well-being of individuals requires that it should be disolved and that the parties should be left free to form such other domestic connections as may contribute to their felicity."[79] Evidently, the use of revolutionary sentiment as a rationale for divorce continued well beyond the end of the American Revolution.

The infant American institution of divorce initiated by Massachusetts Puritans resolved only a portion of the troubled marriages that threatened their vision of a peaceful, orderly society. Judging by abundant advertisements regarding "run-away" wives in the *Boston Evening Post,* deserting one's mate continued to be a common way of terminating marriages during the mid- and late 1700s. The abandoned husband of one Boston women was so chagrined by her departure that he put her name, Lydia, in capital letters in his notice, probably to attract the attention of her relatives and friends. Samuel Perry did the same and more to his wife, Mary. Perry notified the people of Boston that "after repeated acts of infidelity," Mary "hath most wickedly eloped and refused to return to live with me." He concluded, "I do hereby publickly declare that I will not pay one Farthing of any Debtts that she shall contract."[80]

Still, colonial divorce left several important legacies to the nation. Arguments for and against divorce initiated the great American divorce debate of the nineteenth and twentieth centuries. In addition, colonial divorce raised issues that would defy solutions for many years to come: whether to blame one spouse for ending a marriage, what constituted acceptable grounds for divorce, how women were to be supported after divorce, and who was to receive custody of children.

Moreover, the increasing use of divorce to end difficult marriages established an important trend. The growth of colonial divorce was a mere prelude to the proliferation of divorce in the newly established United States.

# Dissolving "Connexions" in the New Nation

☐ After the American Revolution, divorce experienced a rapid expansion in the new United States. By the time Frenchman Michel Chevalier toured the young country during the early 1830s, divorce was already widespread. In his words, marital "connexions" were far more "easily dissolved" in America than in Europe. Chevalier hoped the growth of divorce in America was only a temporary phenomenon, an unfortunate by-product of colonization and westward migration. As a result of this view, Chevalier inaccurately predicted that although Americans left their homes and families "as naturally and with as little emotion as young birds," they would continue to cling to the "sanctity and strictness of the marriage tie."[1]

With historical hindsight, we now know that Chevalier was observing a society in flux; one that continued to believe in marriage as a lifetime agreement, yet was beginning to recognize the need to terminate some marriages by divorce. He also unknowingly was reporting on a formative period in the history of American divorce, the years between 1776, when the newly independent states assumed jurisdiction for divorce, and the early 1850s, when divorce began to emerge as a national scandal. During this period, each region of the new nation—the South, Northeast, and West—embraced divorce with varying degrees of enthusiasm. Increasingly, Americans discussed the nature of marriage, and vociferously debated the desirability of divorce. At the same time, divorce, and laws permitting it, proliferated, and many Americans began to believe that divorce was a citizen's right in a democratic country dedicated to principles of freedom and happiness.

The alteration in attitudes toward divorce was especially noticeable in the South. Southern legislatures, except in South Carolina, made a radical change in their divorce policy after the American

Revolution: although they had opposed absolute divorce during the colonial period, they now began to allow it. Since 1670, in England, Parliament had been granting infrequent divorces to English men and women. After the Revolution, southern legislatures replicated this practice by granting legislative divorces to southern men and women.

In 1790, members of the Maryland General Assembly granted the first divorce in the post-revolutionary South to John Sewall after he had proven to their satisfaction that his wife, Eve, had borne a mulatto child. In subsequent cases, however, wives were more often the wronged parties than husbands. For instance, in 1807, Catherine Dimmett obtained a divorce after she testified that her husband, James, remained in "one continuous state of intoxication" and showed that he was "prepared for the commission of the most desperate and bloody deeds" by repeatedly threatening her life.[2]

By the mid-1830s, the Maryland legislature was granting slightly more than thirty divorces a year. In 1842, legislators, who felt burdened by the volume of divorces, signed a bill placing primary jurisdiction for divorces in courts. Less than a decade later, the Maryland Constitution of 1851 prohibited legislative divorce entirely.[3]

Other southern legislators felt similarly overwhelmed by numbers of divorce petitions as well as by complex charges that demanded extensive investigation. Between 1800 and 1835, for example, divorce-seekers submitted 266 applications to the North Carolina legislature. Over one-third of the petitioners used the clear-cut ground of desertion, often noting that a spouse had relocated "in the western country." The second most common ground, however, was less easily dispatched; almost 8 percent of applicants charged their spouses with committing adultery with African Americans. Other accusations that demanded hearings included adultery with a white person, cruelty, bigamy, impotence, incompatibility, non-support, ill temper, indecent conduct, and "bringing another into house."[4]

Because divorce hearings diverted time and energy from other pressing matters, legislators began to consider shifting jurisdiction for divorce to courts of chancery. In 1806, a Louisiana legislator recommended that the territorial legislature be relieved of "the irksome, disagreeable and laborious task, of tedious investigation" so that it could pursue "objects of greater importance."[5]

It is also conceivable that legislators were beginning to realize that their power and influence lay in enacting laws for broad application rather than in resolving disputes among private claimants. Certainly, during the post-revolutionary years, state legislatures surrendered a wide range of executive and judicial tasks. Legislatures, for example, began to enact general rather than special acts of incorporation. They also gradually surrendered the passing of such special legislation as bills of divorce to chancery courts. At the same time, courts generally seemed willing to become arbiters in interpersonal squabbles, including those between husbands and wives. In fact, common law marriage was itself under increasing attack by chancellors who seemed anxious to regulate citizens' personal behavior.

Yet even as southern legislators began to allow chancery courts to grant divorces, they failed to prohibit state legislatures from continuing to grant divorces. As a result, a confusing, dual system of legislative and judicial divorce existed in most southern states. Only gradually did constitutional provisions and amendments bring a halt to legislative divorce. For instance, North Carolina prohibited it in 1835, Georgia in 1849, and Virginia in 1850.[6]

The course of legislative divorce in Virginia before its termination in 1850 reveals the difficult issues that consumed legislators' time and energies. Legislative divorce, which appeared later in Virginia than in other southern states, soon became a bane to most Virginia legislators. By the early 1800s, it had become apparent that separate maintenance orders—essentially divorces of bed and board—were an inadequate solution to many Virginians' marital problems. Why, for example, should a woman whose husband deserted her be tied to him for life? And what was to be done about the occasional alegal requests for divorce that reached the Virginia General Assembly, especially those that involved heinous marital crimes? In hopes of solving such problems, Virginia legislators began in 1803 to grant acts of divorce to individual petitioners.

As in Maryland, a wife's adultery with a slave man was the volatile issue that persuaded Virginia legislators to enact the state's first "bill of divorcement." In 1802, Dabney Pettus requested a divorce from his wife, Elizabeth. His petition explained that she was "descended from honest and respectable parents" and was supposedly "unsullied in her reputation." Yet "four or five months" after their marriage in March 1801, Elizabeth bore a mulatto child. According to Pettus, she then "publicly and frequently acknowledged" that the child's father was a "negro slave." An 1803 act of divorce returned Elizabeth to *feme sole* status, that of a single woman who could conduct personal business and hold employment, and declared her child "illegitimate & incapable of inheriting any part" of Dabney's estate.[7]

Other cases of interracial adultery soon followed. In 1804, Benjamin Butt requested a divorce because his wife, Lydia, had also supposedly delivered a mulatto child and "publicly acknowledged" that the child's father was a slave. The Butts' 1803 act of divorce returned Lydia to *feme sole* standing, but neglected to clarify the legal position of her child.[8]

The issue of a mulatto child's legal status was a ticklish one that legislators occasionally refused to resolve. On November 25, 1814, Virginia legislators parted Richard and Peggy Jones because Peggy's child was reportedly "the offspring of some man of colour." But because some legislators were unconvinced, the General Assembly directed the Superior Court to impanel a jury to decide whether Richard Jones or a slave man had fathered the child. If the jury ruled that the father was a slave, the child would forfeit its claims on Richard Jones's estate.[9]

Clearly, white-black relationships in this southern, slave-holding society were highly charged issues that could have long-term ramifications. Yet in several legislative divorce cases, a number of women who committed adultery with black men reportedly defied their husbands by continuing their illicit relationships, even bringing their lovers into their own beds. In 1808, a Loudoun County husband testified after discovering that his wife "was of a lewd, incontinent, profligate disposition and practice," he admonished her to cease her adultery. Despite his warnings, he discovered her and "the partner of her crime (a certain James Watts, a man of colour) in bed together." In 1825, another male petitioner stated that in spite of his "remonstrances and persuasions," his wife "has lived for the last six or seven years and still continues to live in open adultery with a negro man, a slave." He added that his wife had borne two children as a result of her relationship with her paramour.[10]

Several Virginia wives who divorced husbands for committing adultery with African American partners (couched in terms of cruelty) also indicated that their husbands ignored their objections to such liaisons. In 1814, an Augusta woman accused her husband of being "criminally, unlawfully, and carnally intimate" with a slave woman named Milly. Testimony revealed that he took Milly into "his own wife's bed and there carried out his licentious designs," and that Milly eventually bore "a mulatto child." A few years later, another woman testified that her husband had taken a slave woman into her bed and had boasted that "he would do it again whenever it suited him." In an 1848 divorce case, a wife complained that she had to sit at the breakfast table with her husband's "slave favorite."[11] Given such high levels of tension and defiance, it is little wonder that such marriages crumbled.

Abuse was another common accusation in Virginia divorce requests. An especially harsh case of abuse was that of Mary Burke. Mary claimed that her husband, Michael, was so idle and intemperate that he destroyed their "domestic happiness." She also testified that he "proceeded from one rash and vicious step to another, until he committed a most wicked and atrocious murder on the body of his brother-in-law, Thomas Warrell." He then fled, leaving her destitute. The General Assembly granted Mary Burke a divorce, but inexplicably ruled that neither party could remarry even though Mary appeared innocent of wrongdoing.[12]

Abandonment was another frequent complaint. In 1819, Mary Brady accused her husband of being "extremely dissipated," committing adultery, and abandoning her and their child. And in an 1842 case, Sally Moffett of Rockingham County testified that her husband, John, had abandoned her. After he left for Ohio ten years earlier, she heard nothing from him. Although the General Assembly granted Moffett a divorce, it temporarily restrained her from remarrying. The Assembly ordered the Rockingham County Court to determine her status. If a jury ruled that John Moffett was legally dead, the court could grant her permission to remarry.[13]

Virginia women were more at risk than men in these divorces because of their financial dependence upon their husbands. Still, although one might expect the General Assembly to be less accessible and more intimidating to Virginia women than to men because of women's exclusion from the political sphere, slightly more legislative divorces went to women than to men between 1803 and 1850. Out of a total of 135 legislative divorces, women obtained sixty-nine divorces, or 51 percent, while men obtained sixty-six divorces, or 49 percent.[14]

Because these women were not all wealthy, many were concerned about their financial prospects after divorce. In legislative divorces, women usually achieved economic independence by returning to *feme sole* standing. Legislative divorce actions routinely stated that a wife's "person and property" were free from control by her former husband. But divorce bills also mandated that men were free of liability for "any promise or contract" subsequently entered into by their former wives; thus, women could not fall back upon former husbands in case of financial need.[15]

If a wife was the guilty party, legislative divorce bills contained an especially strong clause stating that she had forfeited all claims to her husband's earnings and estate. When James M. Martin divorced his wife, Rebecca, in 1842, the legislature declared "all right, interest or claim of the said Rebecca in or to the estate, real or personal, of the said James . . . shall henceforth cease."[16]

Legislative divorce bills were inconsistent concerning the possibility of women remarrying and gaining the financial support of new husbands. During the early years of Virginia legislative divorce, the General Assembly often restrained both parties from remarrying, an action that caused financial hardship to women who had been free of marital error yet were restricted from obtaining the financial support of another husband. Gradually, legislators prohibited only the guilty party from remarrying. In 1820, Barbara W. Pettus divorced her husband, Hugh. Although she was free to remarry, he was barred from remarrying during her lifetime. In a companion action, the General Assembly granted a divorce to Mary Brady and allowed her to remarry, while restraining her husband Thornton from remarrying. This was a seemingly futile action because Thornton had already fled Virginia's jurisdiction; he had "gone to the western country" and "married" another woman while still married to Mary.[17]

The policy of allowing only innocent parties to remarry must have encouraged some women to submit divorce petitions accusing their husbands of error before their husbands charged them with marital crimes. Surely wives preferred having the possibility of remarriage left open to them rather than losing it and its promise of financial support. By the 1840s, however, the Virginia General Assembly virtually stopped prohibiting remarriage, perhaps because such orders were inequitable and usually unenforceable.[18]

Besides worrying about financial support after divorce, many women were afraid that they might lose custody of their children. But legislative divorce records indicate that Virginia women usually received sole control of their children. In a random sample of fifty legislative divorce cases that occurred between 1802 and 1850, women always received custody of children. In 1816, when Anne Cowper divorced her husband William, the legislature deprived him of "power & authority" as "guardian by nature over the persons and property" of his and Anne's children. The act of divorce returned Anne to her *feme sole* status and charged the "proper court" with appointing a guardian for the children. Cowper's children retained their inheritance rights to his estate. Unaccountably, both Anne and William were barred from remarrying even though the legislature evidently saw him as the guilty party.[19]

Subsequent child custody orders omitted mention of legal guardians. When his wife, Nancy, divorced him in 1818, William Anderson lost control over his children, but nothing was said concerning a guardian. As in the Cowper case, the Anderson children retained their inheritance rights to William's estate, and neither spouse was allowed to remarry. A few days later, the General Assembly granted a divorce to Sopha Dobyns, deprived Jonah Dobyns of control of the Dobyns children, omitted mention of a guardian for the children, guaranteed the children's inheritance rights, and prohibited both Sopha and William from remarrying. And when Jane Evans divorced her husband, Ephraim, in 1839, he lost control of the children, but the children retained their inheritance rights to his estate. Nothing was said about a guardian or about the right of remarriage for Jane and Ephraim. These cases indicate that the appointment of a legal guardian was unusual; control of children typically reverted to mothers, a pattern that continued throughout the 1840s.[20]

As their irritation with divorce petitions grew, Virginia legislators began to shift jurisdiction for divorce to Superior Courts of Chancery. In the process, they expanded divorce provisions and put a huge amount of power into the hands of chancellors by charging them to make alimony, child custody, remarriage, and other decisions on the basis of "equity," or fair and reasonable treatment to the parties involved. This situation essentially made divorce a gamble for women because sympathetic chancellors might favor women petitioners in making their awards, while chancellors who believed that women should remain married despite the circumstances would be less inclined to rule for female claimants.

It is likely that legislators were unprepared to assure divorcing women the right to specified financial support, custody of their children, and remarriage for a number of reasons. Such assurances might encourage women to divorce their husbands rather than try to maintain their marriages. Standard legal provisions might also create a class of independent divorced wives and mothers in Virginia society. Finally, guaranteeing women specific legal redress would have given them more power than most early nineteenth-century Virginians were prepared to extend.[21]

The move toward judicial divorce, whose conditions would be based on equity rather than law, began on February 17, 1827, when the Virginia General Assembly legitimized the long-standing chancery court practice of ordering separate maintenance. Chancellors could now grant divorces of bed and board on the grounds of "adultery, cruelty and just cause of bodily fear." The 1827 act also revised the customary practice of granting financial support to aggrieved wives by mandating that chancellors could now "decree to either, out of the property of the other, such maintenance as may be proper; to restore to the injured party, as far as practicable, the rights of property conferred by the marriage, on the other." Chancellors also received the right to dispose of matters of custody and guardianship of children, and provide for financial support of children "as under all the circumstance may seem right."[22]

The 1827 statute also addressed the issue of absolute divorce. It allowed Superior Courts of Chancery to grant divorces on the grounds of "impotency of body at the time of entering into the matrimonial contract, idiocy, and bigamy." In addition, it attempted to regularize legislative divorce by requiring that petitioners file a statement of causes for divorce in the clerk's office of the Superior Court of Law in their home county. This document was to be delivered to a defendant if he or she resided in Virginia; if not, it was to be printed in a Richmond newspaper.[23] This practice, called "notice by publication," was common throughout the United States.

The county court would then impanel a jury to rule upon the alleged causes for divorce. A certified copy of the jury's judgment would accompany every petition presented to the General Assembly. If a divorce from bed and board was previously granted by a Court of Chancery, a record of the proceedings would accompany the divorce petition. This latter provision indicates that the legislature had formally accepted an occasional practice—the conversion of divorces of bed and board into absolute divorces.[24]

The 1827 act also provided both parties the right of appeal in a divorce action. As in other legal suits, either party could take a ruling to the Virginia Court of Appeals. In part, the right of appeal was intended to offer protection to absent defendants, especially those who had failed to see a published notice of an impending divorce suit against them. It gave them a specified period of time in which to appeal a divorce decree handed down in their absence.[25]

In 1841, the General Assembly adopted additional legislation that expanded the list of grounds in chancery court divorces. Grounds were impotency, idiocy, bigamy, and "any other cause for which marriage is annulled by the ecclesiastical law." The act also stated grounds for divorce of bed and board were "adultery, cruelty, just cause of bodily fear, abandonment and desertion, or for any other cause for which a limited divorce is authorized by the principles of the ecclesiastical law," presumably consanguinity and coercion of a minor.[26]

Two years later, additional legislation permitted chancery courts to grant expanded property rights to parties in divorce of bed and board suits. Chancellors could give spouses seeking a divorce of bed and board the same property rights they gave to parties in cases of divorce. In other words, a wife could control her own property rather than receiving payments from her husband. The only difference that remained between divorce of bed and board and absolute divorce was that parties to bed and board cases could not remarry.[27]

Although members of the Virginia General Assembly had attempted to shift a portion of their caseload to courts of chancery, they still felt inundated by divorce petitions. In 1848, legislators grumbled that divorce petitions were "becoming frequent," occupied "much time," and involved "investigations more properly judicial in their nature." Thus, the General Assembly passed an act permitting chancery courts to hear divorce cases involving adultery. Chancellors could also grant the right to remarry to both parties in a divorce or "to the innocent or injured party to marry, and deny it to the guilty party."[28]

In 1850, a new constitution brought legislative divorce to a halt in Virginia. This relieved legislators of an unwelcome task and ended the confusion a dual system of divorce created. But the trend toward expanded divorce continued. Although many Virginians continued to view marriage as an indissolvable lifetime agreement, others argued that spouses had the right to repudiate their marriage vow if it had failed to result in a satisfying marriage.[29]

The Virginia General Assembly appeared to lean toward the pro-divorce position, for after 1850 it continued to expand divorce. Three years after the demise of legislative divorce, the General Assembly adopted legislation listing an enlarged number of marital misdemeanors now considered grounds for divorce: adultery, impotency, confinement in a penitentiary, conviction of an infamous offense prior to marriage, wilful desertion for five years, pregnancy of wife at the time of marriage by person other than the husband, and wife working as a prostitute prior to marriage without the knowledge of the husband. Moreover, divorced people were now free to remarry except in cases of adultery.[30]

As legislators in Virginia—and in other southern states—moved jurisdiction for divorce to chancery courts, numbers of divorce petitions continued to slowly, but steadily increase.[31] This occurred at least in part because courts were more accessible than state legislatures and because chancellors often showed greater flexibility in interpretation than legislators.

In particular, judicial pliancy led to an expanded definition of the ground of cruelty—or indignities, as it was commonly known. Although cruelty was originally defined as "danger to life and limb," judges' rulings broadened it to include inhuman treatment and verbal abuse. In 1832, for example, an Alabama judge softened an 1820 statute that required cruelty must be "life threatening" by ruling that cruelty could also include "inhuman treatment." Other southern judges ruled that cruelty could encompass mental and verbal indignities. In 1849, a member of the Arkansas Supreme Court, Judge Christopher Scott, gave an especially broad construction to the indignities charge by defining it as "rudeness, vulgarity, unmerited reproach, contumely, studied neglect, intentional incivility, injury, manifest disdain, abusive language, malignant ridicule, and every other plain manifestation of settled hate, alienation and estrangement, both of word and action."[32]

Perhaps judges felt compelled to expand the meaning of cruelty as a result of the numerous instances of mistreatment that came to their attention. Charlotte Cullen of Virginia was one such case. In 1844, she wrote to Chancellor Samuel Taylor describing a string of indignities. Cullen explained that although she had adhered to prevailing expectations of women by being "submissive and patient," her husband directed "brutal language" toward her. She said she suffered "mortification & neglect" when her husband circulated falsehoods about her. He also "spurned all her attempts at reconciliation." Cullen reminded Taylor that she was not only a wronged individual, but a wronged woman. "Am I to give up my rights as a wife, a Mother, a Member of Society into the hand of Dr. Cullen, & be driven to commit suicide?" she asked. "If I die my death lies at his door," she continued, "& if the law does not divorce me from him, & stop his persecutions & injustices against me, I will greatly prefer the grave." She pleaded with Taylor to take her case, but it is unknown whether she obtained a divorce.[33]

Some southern wives were partially shielded from such abuse because they had a measure of financial independence from their husbands. Unlike such northern states as Massachusetts and Connecticut, most southern states recognized a married woman's right to hold and administer her own separate family property. Wives were ensured separate estates by pre-nuptial agreements; these contracts were virtually always upheld by chancery courts.[34]

One example of a pre-nuptial contract was an 1824 Virginia agreement between Henry M. Armistead and Mary Robinson. It stipulated that the bride-to-be's "considerable estate in land, slaves, and other personal property" would be held by a trustee for her "sole use and benefit and advantage . . . during her life." This document not only protected Mary Robinson's legacy from being dispersed or squandered by her husband, but guaranteed her financial support should her husband's actions endanger her well-being.[35]

Pre-nuptial contracts were generally utilized by wealthy southern families, but in some states they were negotiated by people of almost all classes. In Louisiana, for example, people of various levels of wealth, racial backgrounds, and ethnic origins signed pre-marital agreements. In 1832, an Irish widow tried to safeguard her plantation and slaves before remarrying. And in 1833, two free African Americans, whose combined estate totaled seven acres, made their marks on a contract protecting each party from debts incurred by the other. Some years later, a free African-American woman, whose dowry consisted of two slaves, also used a contract to protect her property.[36]

Much to the dismay of some families, husbands in need of money, anxious to increase their own wealth, or desirous of undercutting their wives' independence, ignored the provisions of marriage contracts. When Antonia Leonarda Micaela Almonester of Louisiana refused to sign her extensive holdings over to her husband, Celestin Pontalba, he and his father pressured her without respite. Because

Micaela went to court to block their attempted incursion on her property, in 1834 her enraged father-in-law shot her. When she recovered, this strong-willed woman left her husband and managed her considerable property without further interference.[37]

This, and other dramatic cases, reinforces the idea that southern marriages had their share of trouble and conflict, sometimes exacerbated by the presence of large amounts of property. The Myra Clark Gaines case, fought in Louisiana courts and the U. S. Supreme Court between 1835 and 1891, was an especially notorious example of a woman struggling to get title to family property. Because her father had contracted a questionable marriage to a previously married woman, Myra Clark had to seek legal recognition of her status as his legitimate child before she could inherit.[38] Obviously, all was not well with the traditional patriarchal family in the South.

A different situation existed in the Northeast, however. As divorce law developed in the Northeast—the former New England and middle colonies—two important patterns emerged. As in the South, there was a gradual move to eradicate legislative divorce which had become widespread in the years since the end of the Revolution. In Pennsylvania, for instance, a new constitution banned legislative divorce in 1838. An 1849 law did so in Connecticut. The practice of legislative divorce hung on the longest in Delaware, where a new constitution finally terminated it in 1897.[39]

A second important trend was a steady expansion of each state's list of grounds that made divorce easy to obtain for a wide variety of reasons. According to some states' omnibus clauses, a petitioner had only to prove marital breakdown. Unlike the South, this expansion of grounds was often extensive and linked to the growing belief that divorce was a citizen's right in a democratic nation.

Pennsylvania lawmakers initiated what soon became a flurry of divorce legislation in the Northeast. These legislators, who received thirty-five divorce petitions between 1777 and 1785, resented divorce hearings that required enormous amounts of their time. The 1783 case of Elizabeth Keehmle, whose colonial divorce had been disallowed by the Privy Council in 1773, was especially annoying. After almost a year of paperwork and hearings was about to result in a divorce, George Keehmle filed a counter-suit. General Assembly members rejected his application, and in a fit of pique refused to take any further action on Elizabeth's petition. Once again, the Keehmles remained married.[40]

In 1785, the besieged lawmakers freed themselves of such vexations by passing the state's first divorce statute, which contained several important provisions. It established the right of Pennsylvanians to divorce: "where one party is under natural or legal incapacities of faithfully discharging the matrimonial vow, or is guilty of acts and deeds inconsistent with the nature thereof, the laws of every well regulated society ought to give relief to the innocent and injured person." In cases of adultery, bigamy, desertion of four years, and remarriage on the false rumor of the death of a spouse who later reappeared, the injured spouse could go free. The act also placed primary jurisdiction for divorces in the Supreme Court. This move apparently made divorce more accessible, and the process less intimidating, to women, for they had obtained only two of the eleven pre-1785 legislative decrees, but garnered 64 of the 104 judicial decrees granted between 1785 and 1801.[41]

Shortly after the passage of the 1785 divorce law, an *Essay on Marriage* optimistically predicted that Pennsylvania's lenient divorce statute would discourage fraud in courtship, prevent cruelty in marriage, and decrease suicides. After writing that Americans were known for their love of liberty and hatred of tyranny, the anonymous essayist proclaimed a similar "spirit of indulgence" should be extended to "unhappy individuals . . . who are united together in the worst of bondage to each other." It seemed intolerable to this advocate of divorce that the "miserable, henpecked husband, or the abused, and insulted, despised wife" be tied to their unhappy situation until death.[42]

Pennsylvania lawmakers continued to expand the state's list of grounds for divorce. An 1815 law specified cruel and barbarous treatment by the husband and indignities that rendered a wife's condition intolerable as causes for divorce. Legislators added the plea of a wife's lunacy to the list in 1843. In 1854, they recognized two years' imprisonment of either party on a felony charge and a wife's extreme cruelty to a husband as grounds.[43]

Despite a long history of divorce, Massachusetts lawmakers passed a slightly more restrained divorce law than Pennsylvania legislators. In a 1786 act, they shifted primary jurisdiction for divorce from the General Court, composed of the governor and council, to the Supreme Judicial Court (a court of both law and equity) of each county. They also specified the grounds of consanguinity, bigamy, impotency, or adultery by either wife or husband.[44]

As legislators in other northeastern states adopted divorce statutes, they stipulated a variety of grounds. In 1791, New Hampshire lawmakers listed impotency, adultery, extreme cruelty, or three years' absence as grounds for either husbands or wives, and abandonment of three years with failure to provide as a ground for wives. Three years later, New Jersey legislators specified consanguinity, adultery, and desertion of seven years as grounds. And in 1798, Rhode Island officials named impotency, adultery, extreme cruelty, wilful desertion for five years, and a husband's neglect or refusal to provide. In addition, they established what appears to have been the earliest omnibus clause: divorces could be obtained for "gross misbehavior and wickedness in either of the parties, repugnant to and in violation of the marriage covenant."[45]

Lawmakers in new northeastern states enacted similar statutes. For instance, Vermont legislators, who provided for divorce well before statehood became a reality, assigned jurisdiction to Vermont's Supreme Court in a 1779 act. They stipulated as causes for divorce "adultery, or fraudulent contract, or wilful desertion for three years, with total neglect of duty; or in case of seven years' absence of one party, not heard of; after due enquiry is made." Vermont legislators recognized the additional ground of "intolerable severity" in 1787 and added impotency to the list in 1797, six years after Vermont achieved statehood.[46]

Of the northeastern states, Connecticut passed the most comprehensive divorce provisions. Connecticut legislators, for example, added habitual intemperance and intolerable cruelty to an already generous list of grounds in 1843. In 1849, they added life imprisonment, committing an infamous crime, and "any such misconduct as permanently destroys the happiness of the petitioner and defeats the purpose of the marriage relation."[47]

In 1849, the New Haven *Columbian Register* reported several divorce pleas that would easily fall under Connecticut's new omnibus clause concerning misconduct. Polly White complained that her husband "went to bed with his boots on, to annoy her—put dead chickens in her tea pot—would put out the fire when she was up—send her to bed wet and cold." And Sally Beach, married for forty-three years, testified that her husband, Luman, forced her to "sleep in a cold room—had fastened her out of doors—declared he hated her" and "threw a butcher's knife at her."[48]

In that same year, a report in *Niles' National Register* disclosed that a wide range of grounds for divorce existed in the Northeast. In addition to adultery and impotency, *Niles'* reported that northeastern states accepted grounds ranging from desertion to imprisonment, drunkenness, neglect, cruelty, gross misbehavior, and marital "misconduct." New York was the only northeastern state to limit divorce to the sole ground of adultery.[49]

Northeastern legislators were less open-handed in providing for divorcing women. Alimony, for example, was often limited by law. Pennsylvania law was especially restrictive, for alimony could be granted only in divorces of bed and board. Pennsylvania's 1785 divorce statute denied alimony to wives by mandating that "all and every duties, rights and claims accruing to either" spouse ceased with divorce. Connecticut divorce law allowed wives a portion of joint property, usually one-third to one-half of a couple's estate. And in New York, a wife who could prove that her husband committed adultery, while her own behavior was above reproach, might regain part or all of the property that she had brought to the marriage and receive alimony as well.[50]

Divorce in the Northeast, then, was similar to southern divorce in that legislators gradually shifted jurisdiction for divorce from their own hands to those of judges. Unlike southern legislators, however, northeastern lawmakers tended to establish more inclusive lists of grounds for judicial divorce, while being more hesitant to rely on judges' ability to apply standards of equity in making alimony awards and other decisions.

As the rapidly growing population of the United States began to spill over the Appalachian Mountains and settle frontier regions between the Appalachians and the Mississippi River, the institution of divorce took on additional flexibility and even an air of casualness. Because westward migration involved huge distances and rudimentary communications, marital desertion was especially common and often led to divorce. Men went west, some believing that they would eventually reunite with their wives, others knowing that they would not. Women refused to join their migrating husbands or left their western-based husbands and returned to their former homes.

When Alexis de Tocqueville related his impressions of the United States in 1838, he declared that "there is certainly no country in the world where the tie of marriage is so much respected as in America, or where conjugal happiness is more highly or worthily appreciated." But Tocqueville also noted the existence of "a restless disposition" and an "excessive love of independence" among Americans. Moreover, he remarked that continuous migration toward westward regions kept "ties broken or unformed."[51]

Western settlers soon formalized what Tocqueville called broken ties by enacting divorce statutes bearing marked similarity to those of northeastern and southern states. In Iowa, for example, the expansion of grounds followed a pattern very like that of northeastern states. When Iowa became a territory in 1838, legislators specified impotency and adultery as grounds for divorce. Iowa's Revised Statutes of 1842–43 listed eight grounds for divorce: impotency, bigamy, adultery, desertion for one year, conviction of a felony or infamous crime, habitual drunkenness, cruel and barbarous treatment endangering the life of spouse, and indignities rendering a spouse's situation intolerable. In 1845, an amendment added a ninth cause—an omnibus clause that permitted judges to grant divorces "when it shall be made fully apparent to the satisfaction of the court, that the parties cannot live in peace or happiness together, and that their welfare requires a separation between them."[52] When Iowa achieved statehood in 1846, these provisions remained in force and unaltered until 1851.

In another area, early Iowa divorce legislation more closely resembled southern policy by placing extensive discretionary power in the hands of judges. In 1838–39, the Iowa Code stated that in divorce cases based on a wife's adultery "the husband shall have the personal estate for ever, and the real estate of the wife," but a court could "allow for her subsistence" out of the couple's holdings as it saw fit. If the wife was the innocent party, she was to receive the property she owned at the time of marriage and could expect the court to order alimony payments up to one-half of the husband's income. The Code of 1842–43 put such decisions into the hands of judges by ruling that courts would determine both property matters and alimony payments. Judges were still expected to restore an innocent wife's property to her and to grant alimony payment to wives who were innocent parties.[53]

Subsequent legislation reaffirmed judicial discretion. The Iowa Code of 1851 stated that the court could make decisions regarding the "property of the parties and the maintenance of the wife as shall be right and proper." Under this rule, courts could, and did, give a share of a husband's property to a wife as alimony, thus allowing her more freedom to manage her own affairs than she would have with payments. In addition, judges sometimes awarded guilty wives property settlements and alimony awards.[54]

In other instances, however, western lawmakers not only borrowed, but also revised, older divorce legislation. For instance, when Illinois legislators passed an 1825 divorce statute, they specified the grounds of impotency, adultery, and two years' absence as grounds for divorce, prohibited the guilty spouse from remarrying, and permitted divorces of bed and board. But Illinois lawmakers added their own innovations to the statute. Rather than restraining a guilty spouse from remarrying for life, the 1825 law prohibited remarriage for two years. Moreover, divorces of bed and board could be obtained *by either party* for cruelty or intemperance.[55]

Similarly, western legislators followed—and enlarged—an occasional practice of southern and northeastern legislatures. Like some of their counterparts in older states, western lawmakers often dealt with a glut of business by divorcing more than one couple at a time. An Illinois act of 1825 divorced both Thomas and Eliza Shannon of White County and Matthias and Martha Drain of Clinton County from "the bonds of matrimony." Other western states expanded this practice dramatically; in 1842, the Iowa territorial legislature reportedly divorced a total of eighteen couples in one bill.[56]

Western divorce laws, then, were neither totally imitative nor wildly innovative. But they were more widely utilized than in the other two regions. Between 1776 and 1850, western states and territories surpassed both the Northeast and the South in numbers of divorces granted.[57] During the latter half of the nineteenth century, divorce in the American West, more fully discussed in Chapter 4, expanded to a point that many Americans found distressing and alarming.

## Women and Divorce

Recently, numerous scholars have analyzed the alterations that occurred in the lives of American women between 1776 and 1850, especially the emergence of a separate "woman's sphere"—meaning that women were to operate within the home, while men operated in realms outside the home. Yet no one has studied the impact of separate spheres on marriage and divorce. One might theorize that the separation of women's and men's spheres into different areas of influence relieved nineteenth-century marriages of stress. By giving wives and husbands their own distinct domains to rule, the emergence of separate spheres may have reduced conflict between mates.

One might also speculate that the concept of separate spheres increased pressures on marriage. Certainly, women taught to view themselves as private actors and men taught to think of themselves as public influences would bring radically different world views, values, and ways of expressing affection to marriage, thus increasing the differences between wives and husbands. In addition, such women and men would probably bring raised—and often unrealistic—expectations to marriage. A wife who expected to remain within the home yet discovered that she had to take employment outside of the home might be distressed indeed. And the husband who expected to be free of domestic duties yet found that he had to help an unprepared or overburdened wife with domestic affairs might be disillusioned with his spouse in particular and marriage in general.

Other issues that remain unclear, despite extensive study of the impact of rapidly changing divorce statutes and practices, is the effect of divorce, alimony, and child custody decisions on women.[58] Certainly, individual instances show that divorce could be extremely hurtful to women.

An especially poignant case was that of Lavina Davison who married Jacob E. Whisler in Indiana on October 23, 1836. The Whislers soon relocated in Tipton County, where Lavina bore their first child on August 26, 1837, and bore three more children, in 1838, 1840, and 1842. In May 1845, Jacob applied for a divorce, claiming that

Lavina was insane, unable to carry on a sexual relationship, and had been living with her mother for three years. One can only wonder if Lavina sought refuge with her mother to escape constant childbearing and Jacob. Obviously, Jacob had little commitment to marriage as a lifetime agreement, for he quickly abandoned his wife when she became useless to him. After divorcing Lavina in 1845, he remarried in April 1846.[59]

The episode was far from over for Lavina. Jacob obtained custody of their four children, and Lavina was denied alimony although her mother received a small sum for Lavina's care. In 1850, a jury of twelve men declared that Lavina was insane. The court placed her under the guardianship of an Andrew J. Redmon. Because she was uneducated, Lavina signed the court order with an X.[60]

Lavina's story has two possible interpretations. Perhaps the legal system was working for Lavina in particular and society in general by protecting and sustaining her in the face of her own inability to do so. The belief of some family members that she died in the Tipton Infirmary reinforces the view that she was an incapable woman who was well cared for by the legal system. Or, perhaps the legal system was working against Lavina by allowing her husband to leave her in a position of financial and psychological dependence upon others. The belief of other family members that Lavina eventually "got better," remarried, and left Indiana, supports the view that Lavina was a healthy woman, temporarily weakened by Jacob and a male court.

What the Whisler records do clearly reveal is that courts could easily strip women of the financial support they had earned as wives. Alimony, usually in the form of cash payments or property, was to replace the financial support pledged by a husband at the time of marriage if his wife had not forfeited it by committing a marital crime. Because most wives were financially dependent upon their husbands, they needed some type of alimony to survive after divorce.

Divergent laws and practices make it difficult to formulate generalizations about alimony. Certainly, the widespread custom of awarding alimony primarily to innocent wives must have encouraged many wives to seek a divorce before their husbands did so. And husbands determined to avoid alimony must have flung many countercharges against wives who were trying to divorce them. Because divorce cases involving alimony clearly had high stakes, they exacerbated the adversarial nature of divorce proceedings.

In addition, because alimony was uncertain, many wives must have requested a divorce of bed and board with its virtual guarantee of support rather than asking for a divorce. The disadvantage of a divorce of bed and board was that the husband retained control of the couple's property. A wife had little recourse if her husband mismanaged their property or refused to pay alimony. She was unable to act on her own behalf because, as a separated rather than a divorced woman, her *feme covert* status prohibited her from running a business, signing contracts, and controlling property. Nor could she remarry during her husband's lifetime.

A woman who obtained a divorce could expect a different outcome. She might regain her dowry and a portion of the couple's property, and would achieve economic independence by returning to *feme sole* status. If she was the innocent party, she could remarry, thus gaining a husband's financial support.

In many states, judges were more generous with temporary alimony than with permanent alimony. Temporary alimony was to pay a wife's living and legal expenses while litigation was pending. The number of temporary alimony awards increased as more women filed for divorce.[61] Undoubtedly, temporary alimony enabled women of little means to pursue divorces otherwise beyond their financial reach.

Child custody was another controversial and highly emotional issue. Because many women feared losing their children in divorce actions, they remained in destructive marriages for years. In 1783, the wife of a dissolute, abusive man explained that she stuck with her husband because she feared that he would gain custody of her "dear infant" in a divorce suit. Although she believed it was impossible for her "to live happily with him," she wanted to try. "What wou'd I not bear to be able to keep my Child with me?" she noted in her journal.[62]

English-born actress Frances Kemble was another example of a woman who stayed married largely to avoid losing her children. After years of arguments with her husband, Georgia planter Pierce Butler, regarding the management of their Philadelphia home, their two daughters, his ownership of slaves, and her desire to return to the stage, she sailed for England in 1845 to perform in London theaters. Butler divorced Kemble for desertion in 1849. The decree gave her a yearly allowance of $1500 and two months' visitation per year with her children. Although Kemble regained her personal, social, and economic independence, she essentially lost her children in the process.[63]

Distressed wives concocted a variety of solutions to prevent such loss of children. In 1830, a distraught woman related a violent episode to her sister. Her husband had plunged a knife into her ribs, but a corset bone had stopped the blade. She implored her sister to take her daughter, Eugenia, into her home. Although giving Eugenia away was like "tearing" flesh from her own body, she felt unable to raise the girl "as she ought to be under the circumstances." Neither could she ensure that she would retain control of Eugenia in the event of divorce.[64]

Such heart-wrenching tales elicit sympathy for women who lost custody of their children, but there is another side to the story. The idealization of motherhood after the American Revolution encouraged many judges to reject the traditional notion that children belonged to fathers. By the 1820s, many judges had adopted the newer view that young children were better off with their mothers.[65]

Judges also considered the matter of guilt in making child custody decisions. If a female petitioner could prove her husband's guilt to a judge's satisfaction, she was likely to receive custody of their children. Some judges believed that young children should be raised by their mother even if she was the guilty party in a divorce. For instance, in 1813, Pennsylvania Chief Justice William Tilghman awarded Barbara Lee the custody of her two daughters after her husband divorced her on the ground of adultery. Tilghman explained that, "considering their tender years," the two girls needed the kind of care "which can be afforded by none so well as a mother."[66]

The question of the effects of divorce on women will have to be resolved through state-by-state research.[67] It is true, for example, that New York and Louisiana divorce laws restricted grounds for divorce and established procedures that made it difficult for women to obtain divorces. On one occasion, Governor Thomas B. Robertson of Louisiana flatly refused to approve five divorces, maintaining the "frequency of divorce is destructive of happiness and virtue" in American society. But, in such other states as Connecticut and Tennessee, the growing numbers of female petitioners and the improvements in their success rate dispute the belief that divorce was punitive to women and difficult for them to obtain.[68]

Even in strict jurisdictions, American women had more legal recourse than their mothers and grandmothers had. Moreover, divorce gave American women legal relief that was out of the reach of women in other countries. In England, divorce was virtually unavailable to women. Not until 1801 did a woman petition the House of Lords for a divorce. In that year, Jane Addison obtained a Parliamentary divorce because of her husband's adultery. Despite Addison's breakthrough, only four English women obtained divorces by 1850 even though Parliamentary divorce had existed for 186 years.[69]

When the English author Harriet Martineau toured the United States during the mid-1830s, she was struck by the relative ease of divorce in America. She lamented English practices that limited divorce to "the very rich." Martineau especially commended Massachusetts authorities for defining the ground of cruelty in a broad way. During a visit to North Carolina, she came to believe that divorce was accessible even to impecunious women. There, she met the wife of a "gamester" who obtained a divorce "without the slightest difficulty."

According to Martineau, when the wronged wife "brought evidence of the danger to herself and her children,—danger pecuniary and moral,—from her husband's gambling habits, the bill passed both Houses without a dissenting voice."[70]

Still, divorce was an imperfect institution that contained far more dangers for women than for men in the new United States. Divorcing men often divested themselves of wives, children, and financial obligations through divorce. Even when judges ordered men to pay alimony, such an order was difficult to enforce. Thus, the default rate was high. Because men were generally financially independent, they simply continued to support themselves after marriage, while financially dependent women were unable to do so. Although women in the fledgling nation had legal options unavailable to most other women, they had to utilize the option of divorce carefully.

## Changing Expectations

Despite the many pitfalls in divorce, there is no doubt that after 1776 a growing number of people who had promised to be "true & faithful" until death, abandoned their spouses by way of the divorce court.[71] Although it is impossible to calculate a formal divorce rate for the new nation because state officials neglected to collect statistics regarding marriages and divorces during the early years of the new nation, scattered statistics indicate that divorce was gaining in popularity. In Massachusetts, divorce pleas by both wives and husbands increased during the decade following independence, while in Norwich, Connecticut, fifteen people requested divorces between 1779 and 1799.[72]

To some Americans, divorce seemed to be getting out of hand. In 1788, Congregational minister Benjamin Trumbull expressed outrage that 390 couples had received divorces in Connecticut during the preceding fifty years. He strongly urged officials to follow the biblical injunction to grant divorces only in cases of adultery.[73]

Others, however, took the number of divorces in stride. Connecticut legislators continued to broaden divorce provisions rather than implement restrictive measures as Trumbull and other ministers recommended. In 1795, Judge Zephaniah Swift complimented Connecticut policy-makers on what he saw as enlightened action. Swift viewed Connecticut's divorce policies as "temperate" and conducive "to the virtue and happiness of mankind."[74]

Of course, these commentators were primarily referring to the phenomenon of divorce among white Protestant Americans. The appearances of an American Indian petitioner and a Roman Catholic divorce-seeker in Pennsylvania courts in 1795 were highly unusual events.[75] African American petitioners also appeared only rarely. Although divorce petitions sometimes named African American men and women as co-respondents in adultery cases, African Americans seldom sued for divorce. Because most African Americans were slaves, the white legal system refused to recognize their marriages, much less terminate them. Instead, slaves themselves, their owners, or their churches made decisions about their marriages, separations, and divorces.[76]

Commentators also saw only the proverbial tip of the iceberg of marital unhappiness: many more people were unhappy than sought divorces; many more people applied for divorces than obtained them. Between 1777 and 1785, the Pennsylvania legislature investigated thirty-five appeals and granted eleven divorces. In 1810, members of the North Carolina General Assembly considered twenty petitions and granted one divorce. Three years later, they received twenty-two applications and granted four decrees. In Tennessee, the legislature received 426 petitions between 1797 and 1833, and granted 111 divorces.[77]

Still other divorce pleas were omitted from judicial records because divorce-seekers directed them to inappropriate people. The 1789 register of Philadelphia's Gloria Dei Church (Lutheran) reveals that embattled couples sometimes requested ministers to "divorce" them. Some clergy complied, even assuring mates that they could lawfully remarry, but others refused to do so. A Philadelphia minister tersely described his refusal to part an unhappy couple in 1795: "He gets drunk and beats the wife. She seems to have a cutting tongue. Dismissed with proper advice."[78]

Numerous couples ended their marriages in ways other than divorce. In South Carolina, where divorce was prohibited, and in neighboring North Carolina, where divorces could be obtained only for impotency or adultery before 1827, numerous wives sought divorces of bed and board. Still other North Carolinians negotiated their own separation contracts and even agreed that both parties could remarry. In 1803, John and Rebekah Farrow signed a document stating that because "unhappy Differences" existed between them, they "Mutually agreed to separate themselves." They also agreed that each was free to marry without fear of "prosecution" from the other, and that "all children hereafter Begotten by Either . . . shall not in any wise stand chargeable to the adverce party."[79]

Yet other marriages ended in desertion. Newspaper notices of abandonments and "runaway" spouses show that many people simply left their marriages behind them. One instance was that of Reuben Warson of Howard County, Missouri. In 1824, Warson placed a newspaper notice stating he thought it "advisable" that he and his wife "tear" themselves apart for "just causes and provocations."[80] In today's world of Social Security numbers and computer records, desertion is somewhat more difficult; departing spouses often get divorces, become part of the statistical record, and swell the divorce rate. Then, however, thousands of desertion cases escaped inclusion in formal records.

By 1850, it was clear that the time-honored pledge of "true & faithful until death" had little long-term value to many people who had recited it to their spouses on their wedding days. Numerous factors, including industrialization, urbanization, westward migration, and changing roles of women, caused marriages to come apart at the seams. Industrialization, for example, supplied a growing number of Americans with the financial resources to divorce. In addition, the rise of an American business community demanded legal clarification of property titles and business interests, which in turn called for a relatively simple divorce mechanism so that people would divorce rather than desert and leave unclear property rights in their wake.[81]

Another stress factor was Americans' rising expectations of marriage, which created more disappointment with marriage—and thus more divorce. After the American Revolution, the customary view of marriage as a patriarchal structure was increasingly challenged by an emerging ideal of companionate marriage—a union based on a partnership of friends and equals. In addition to usual expectations that spouses would establish a sexual relationship, have children, and be economic partners, Americans stressed more and more the growing importance of three qualities in marriage: respect, reciprocity, and romance.

Spurred on by revolutionary rhetoric against submission and tyranny, husbands and wives began to ask for increased respect from their spouses. Of course, wives had long been taught to respect their husbands, but they now began to ask for more honor, esteem, and consideration for themselves. They also questioned the traditional idea that wives should be submissive, especially to husbands who were thoughtless and abusive. As early as 1776, political wife and farm manager Abigail Adams objected to the "tyranny" that husbands exercised over wives. In her view, wives should be helpmeets rather than slaves.[82]

Growing numbers of female divorce-seekers also objected to spousal tyranny. In an 1805 plea, a rebellious wife who had left her alcoholic, abusive husband testified "she did not find it her duty to struggle any longer with her afflictions."[83] This woman rejected the customary prescription that a wife be submissive. She decided that she owed allegiance to herself and her six children rather than to a husband who had forfeited it by his actions. To her, the idea of marriage as a lifetime contract was clearly inapplicable.

Popular literature also exhibited Americans' growing concern with increased respect in marriage. In 1792, a woman who styled herself a "Matrimonial Republican" objected to the word "obey" in marriage ceremonies because it connoted master-slave relationships. She asserted that marriage should mean a "union of interest" that resulted in a "partnership." And in 1831, an essay in *Godey's Lady's Book* announced that tyranny was "out of fashion" and assured its readers that "men are becoming more enlightened and more rational."[84]

An 1848 story in *Godey's* demonstrated what could occur when a man refused to be enlightened and rational. Author Grace Greenwood described a woman whose fiancé asked her to give up her career as a "reputed poetess." Greenwood's heroine replied that this was too much to ask: "I could only promise to love you through all life, with the proud love of freedom and equality; a love which, trust me, is feminine in the voluntary homage of conscious strength." When he failed to see her point, he lost his future wife: she refused to marry him.[85]

In 1852, writer T. S. Arthur presented another hard case. Because Arthur wrote primarily for a female audience, he discussed issues of importance to women and presented a view attractive to them. In this novel, Arthur's protagonist was a pompous man named Henry who regarded women as inferior. His attempts to rule his wife eventually drove her away. They were reconciled only after Henry began to regard her as an equal and to relate to her on the basis of "mutual deference, confidence," and "respect."[86]

Women's rights' reformers similarly called for respect for wives. In her 1838 books, *Letters on the Equality of the Sexes*, abolitionist and feminist Sarah Grimké claimed God intended wives to be "companions, equals, and helpers" rather than housekeepers. "Men and women were CREATED EQUAL; they are both moral and accountable beings," she insisted. Like Adams and "A Matrimonial Republican," Grimké employed the slave metaphor: women must "rise up from slavehood" and become respected partners of men. In 1843 and again in 1845, Transcendentalist Margaret Fuller also defined wives as companions: equal parties to "household partnerships."[87]

In addition to heightened respect, some Americans were beginning to see reciprocity as an important ingredient in marriage. One manifestation of this attitude was the growing number of wives who were unwilling to accept less than reciprocal fidelity: they divorced their husbands for adultery. In Connecticut between 1789 and 1793, over one-fourth of female petitioners cited their husbands' adultery as grounds for divorce. These women apparently expected faithfulness to be a two-way agreement and resorted to divorce when they discovered their husbands were unfaithful.[88]

Popular writers reflected the growing concern with increased reciprocity in marriage by discussing its importance at length. As early as 1784, one writer explained that marital "felicity" meant that a wife and husband tried to please each other whenever possible. T. S. Arthur's widely read *Sweethearts and Wives; or, Before and After Marriage* especially argued for reciprocity in marriage. Arthur's didactic tale of a wilful husband and prideful wife who almost destroyed their marriage ended with an unmistakable moral: reciprocity had to exist if a marriage was to work.[89]

Some Americans tried to establish reciprocity in their marriages by signing marriage contracts. When reformer Robert Dale Owen married Mary Robinson in 1832, he signed a statement that would have pleased the "Matrimonial Republican." Owen repudiated his legal powers over Mary, explaining that although it was impossible for him legally to surrender "the unjust rights" which "an iniquitous law tacitly give over the person and property of another," he could do so morally. To Owen, a husband's rights were the "barbarous relics of a feudal, despotic system." In subsequent years, a handful of other couples, notably Lucy Stone and Henry Blackwell, signed similar contracts in hopes of creating marriages based on mutual duties and expectations rather than on husbandly power and wifely subservience.[90]

The third emerging component of marital happiness in the new nation was heightened romantic love. Letters written by courting couples exhibited an intensified concern with romantic feelings rather than practical matters and disclosed that a growing number of couples began to reject parental advice regarding their choice of mates. Instead, they based their decisions to wed on the presence of love.[91]

Divorce petitions demonstrated that some Americans expected the romantic love they had known as courting couples to continue after marriage. A Connecticut marriage fell apart in 1777 because, according to Amariah Kibbee, "tho' melancholly are the Truth," he no longer loved his wife enough to oppose her divorce petition. During the 1780s, approximately one-tenth of Massachusetts divorce petitions mentioned love. Such terms as "alienated affection," "nuptial happiness," and "broken heart" became common. The lack of love at the time of marriage caused problems as well. An especially pitiful case occurred in 1833 when a seventeen-year-old North Carolina woman requested a divorce from a man she did not "favour." Her father, who had pushed her into the loveless marriage, saw the error of his ways and aided her in obtaining a divorce.[92]

Romantic love also provided a theme for thousands of essays, stories, and novels between 1776 and 1850. A 1785 essay in the *Boston Magazine* encouraged young people to resist the efforts of ill-advised parents to coerce them into loveless marriages. Although they should consider their parents' judgment, they must "avoid sacrificing a life of happiness" by bending to parental will. In an 1831 story in *Godey's Lady's Book,* a domineering father was forced to recognize his pushiness at the last possible minute. To his credit, he replaced his handpicked groom with the one his daughter loved.[93]

T. S. Arthur also counseled couples to marry for love. In his 1843 novel, *The Stolen Wife: An American Romance,* Arthur described the fortunes of a young woman whose true love carried her off a scant half-hour before she was to wed a man of her parents' choice. In another of his popular works, Arthur portrayed a young woman who held strong views about love. Arthur's heroine, Flora, believed in love as the sole basis of marriage, thus she refused two proposals from men she did not love. In an era when most women married, she audaciously stated that she would rather remain single than contract a loveless union.[94]

The qualities of respect, reciprocity, and romance increasingly gained popularity, but they failed to gain total approval. Although discussion of the "three R's" became widespread during the early nineteenth century, other people reiterated traditional views. Repeatedly, women were advised to behave in a complacent, agreeable way. Wives were counseled to be submissive, passive, and obedient and to shine in their "proper sphere"—the home.[95]

As wives, women were to find meaning in life by caring for their husbands, offering husbands solace, counseling, and renewal after they had expended their "whole moral force in the conflicts of the world." To be effective wives, women were not only to remain in their proper sphere, but were "to live in the regions of sentiments and imagination."[96]

Woman's sphere also included motherhood. In the new United States, a growing idealization of motherhood resulted in reverence for one mother in particular—Mary Ball Washington (George Washington's mother)—and for republican mothers in general—mothers who raised virtuous American men and women. In 1838, novelist Lydia Sigourney's *Letters to Mothers* summed up motherhood this way: "Hath any being on earth a charge more fearfully important than that of the Mother?" And in 1842, writer Margaret Coxe asserted that the "maternal relation" was the "most important channel through which woman was to direct her special moral agency."[97]

As a result of the emergence of the three R's alongside these traditional views, a dual system of values soon existed, and confusion frequently reigned in romantic relationships. Nathaniel Hawthorne's 1850 novel, *The Scarlet Letter,* gave one person's view of the situation. Hawthorne described throngs of distressed women who came to an older and wiser Pearl for advice. Made wretched by their attempts to achieve the elusive ideal of a happy marriage, they demanded relief from their despair. After calming each as best she could, Pearl predicted relationships would be easier in the future: "at some brighter period, when the world should have grown ripe for it, in Heaven's own time, a new truth would be revealed, in order to establish the whole relation between man and woman on a surer ground of mutual happiness."[98]

It seems likely that growing expectations concerning happiness in marriage had themselves become disruptive forces in numerous American marriages.

## The Great American Divorce Debate

After the American Revolution, Americans not only discussed marriage at length, but fervently debated divorce as well. Conflicting views spewed forth from pulpits and other platforms, filled columns in newspapers, and prompted the publication of early divorce stories throughout the United States. The intensity of the American divorce debates reflected the issue's tremendous volatility.

There is no doubt, of course, that divorce was a complicated matter. Although legal divorce was a relatively straight-forward procedure that simply voided a marriage contract when one party violated it by committing adultery, bigamy, or behaving in some other unacceptable way, it was seldom this simple when applied to particular couples. Most divorce suits involved far more perplexing moral and emotional issues than did the usual contract termination.

Opponents of divorce believed that the spread of divorce revealed decay in American society and breakdown in the American family. Some years after Congregational minister Benjamin Trumbull delivered his diatribe against divorce in 1788, Yale President Timothy Dwight similarly condemned divorce. According to Dwight, "The progress of this evil" was "alarming and terrible." Noting that one out of every hundred Connecticut couples had divorced in recent years, Dwight exclaimed: "What a flaming proof is here of the baleful influence of this corruption on a people, otherwise remarkably distinguished for their intelligence, morals, and religion!"[99]

Other Americans who opposed divorce believed it was justified in certain cases. Representative Samuel S. Mahon of Concordia County, Orleans Territory was a well-educated, articulate member of the House of Representatives and its "committee on divorce and alimony." In 1806, he spoke heatedly against what he termed "the incipient evil of divorce." In his view, divorce contributed the "lowest ebb of degeneracy" to society, undermined the "fundamental principles of society itself," and directly opposed the "fundamental principles of matrimony itself." But Mahon also noted divorce provided a solution to difficult marriages, especially those that involved impotency, bigamy, and consanguinity. Mahon's contradictory stance was far from atypical: although he feared for the future of the American family and believed that marriage should last a lifetime, he also opened the door more than a crack to divorce.[100]

Other Americans took a far more radical view of divorce. Utopian reformers especially viewed marriage as a voidable agreement and thus worked energetically on behalf of divorce. Reformer Robert Dale Owen dramatically chose Independence Day—July 4, 1826—to declare that no one should be forced to remain in a distressing marriage. Two years later, reformer Frances Wright presented an even more extreme attitude in the *Memphis Advocate*. Wright disclosed that marriage was non-existent in her Tennessee commune, Nashoba, established as part of her plan for the eventual emancipation of slaves. The purpose was to keep women from forfeiting their "individual rights or independent existence." The absence of marriage would also prevent men from asserting "any rights or power" over women beyond what they exercised over women's "free and voluntary affections." Offended Americans responded to Wright by branding Nashoba a "free love colony."[101]

As the divorce debate captured more newspaper space, divorce became a theme for popular writers as well. In 1841, Emma Embury was one of the first to deal with divorce. In "The Mistaken Choice," she described a despairing young husband whose wife's actions led him into bankruptcy. After his wife sailed to France with a mysterious, bearded stranger, he divorced her and began over as a lowly clerk.[102]

In 1850, two early divorce novels appeared, but their authors were far less favorable to divorce than Embury. One, by revered female novelist E.D.E.N. Southworth, was *The Deserted Wife*. In a foreword, Southworth railed against divorce: "in no other civilized country in the world is marriage contracted, or dissolved, with such culpable levity as our own." Nowhere else could "divorce be obtained with such facility, and upon such slight grounds." Although it was nigh impossible for a man to divorce a faithful wife, he could "oblige her to release him, or break her neck, or her heart, or desert and starve her into compliance with his measures." Or, Southworth added, he could "wrest her children from her, and make their restoration to her bosom the price of his release." The answer to marital conflict, she insisted, lay not in easy divorce, but in better preparation for marriage.[103]

Southworth's tale supported her argument. The heroine, an immature, jealousy-prone wife, soon alienated her weak-willed and self-indulgent husband. When he and his money disappeared somewhere in Europe, she pulled herself together, as Southworth's women were wont to do. She cared for their three children, became a concert singer, and amassed an impressive estate. When he reappeared, she, being of "high heart," reunited with him in "love and joy." Southworth concluded that better preparation for marriage would have saved this couple considerable grief and could save many needless divorces as well.[104]

T. S. Arthur's 1850 tale, *The Divorced Wife,* was another of the early divorce novels. Believing scurrilous stories about his wife, Arthur's protagonist obtained a divorce and custody of the couple's two children. Aided by a loyal servant, his wife kidnapped their daughter and hid her in a rural retreat. After two years of frenzied searching, the husband located them. Upon seeing his wife's love for their daughter, he realized that he had falsely accused her. He began to visit his former wife and their daughter in their humble cottage. After three months, the couple remarried, returned to their fine home, and "lived as a family happily."[105] As in Southworth's story, preparation for marriage rather than divorce would have provided an answer to this couple's problems.

As reformers, novelists, and other concerned people raised questions, explored the meaning of divorce, and frequently railed against divorce, they escalated the divorce debate in America. Although this dialogue aired issues, it also interfered with their resolution. Conflict between supporters and opponents of divorce diverted attention from the idea of creating laws, procedures, and policies that would help divorcing mates through the process.

In 1850, what Michel Chevalier had called "family sentiment" remained strong in the United States, but the ease of dissolving "connexions" had increased markedly. Divorce was now firmly entrenched in the laws of most states and territories. Also, growing numbers of Americans had accepted the concept of divorce as a right: that people suffering abuse or tyranny at the hands of mates were justified in seeking their personal freedom. It was perhaps a sign of the times that after Tennessean Andrew Jackson married a divorced woman, Rachel Donelson Robards, and did so under scandalous circumstances, the nation still elevated him to the presidency of the United States in 1828 and 1832.[106]

Still, even though divorce had woven itself into the very fabric of society of the new nation, many Americans continued to oppose it. By 1850, the impassioned conflict between the pro-divorce and anti-divorce factions was tottering on the brink of becoming a major controversy. The groundwork was in place for an acceleration of the great American divorce debate and for an extraordinary expansion of divorce into a national scandal.

# "Unmarried
Nearly at
Pleasure"

☐ Editor Horace Greeley was outraged. In 1852, he vented his ire in the *New York Tribune* by warning Americans that individualism was running amok. "This is preeminently an age of Individualism," Greeley complained. "The right of every man to do pretty nearly as he pleases" is "gaining ground daily." To Greeley, divorce was the single most destructive manifestation of individualism. In his view, divorce would soon result in "a general profligacy and corruption such as this country has never known."[1]

A few years later, Greeley launched another vitriolic attack on divorce. On this occasion, he was upset by lenient divorce statutes and sloppy divorce procedures in Indiana. In a passionate outburst, he branded the Hoosier state "the paradise of free-lovers," where people could "get unmarried nearly at pleasure."[2]

Greeley's harsh statements concerning divorce reflect the growing intensity of the divorce debate in the United States during the Civil War era. Between 1850 and the beginning of a divorce reform movement around 1880, divorce was frequently in the spotlight. Public opinion was regularly inflamed by reports of a divorce mill in Indiana and of radical-sounding experiments with divorce by utopian communities in upstate New York. Only South Carolina remained inviolate. With growing fury, opponents railed against divorce, while supporters argued on its behalf. In the meantime, the divorce rate continued to climb. Although Americans held fast to their belief in love and marriage, it was apparent that growing numbers of them saw their marriage vows as a dissolvable contract.

Before 1850, Connecticut, Rhode Island, Vermont, Maine, Pennsylvania, and Ohio allegedly granted divorces with a liberal hand to their own citizens and to migratory divorce-seekers from states with more stringent divorce laws.[3] After 1850, however, Indiana became a forerunner to Reno when the state's divorce laws reportedly attracted huge numbers of migratory divorce-seekers. Public alarm became evident as dramatic reports described the Hoosier state as a divorce mecca, churning out easy divorces to people from stricter states with little regard for long-term consequences to spouses and children.

Indiana's rise to notoriety began unobtrusively in October 1850, when delegates, intent on drafting a new state constitution, found themselves embroiled in heated debate over divorce. Prior to 1850, both the Indiana legislature and county courts had granted divorces, but legislators had asked to be relieved of these time-consuming duties. Faced with this pressure to revise Indiana's system, convention members finally agreed to abolish legislative divorce and to give courts exclusive jurisdiction in divorce suits.[4]

As some convention delegates had feared, this move set the stage for an expansion of judicial power. In 1852, a legislative committee, reportedly under the leadership of reformer and divorce advocate Robert Dale Owen, formulated a list of grounds for divorce. Committee members agreed to retain the state's existing eight grounds that included several typical clauses and a broad omnibus clause passed in 1824. This catch-all omnibus clause allowed judges to grant divorces for grounds other than those on the list if they thought them "proper" causes for divorce. When Horace Greeley attacked the looseness of the omnibus clause, Owen replied that it was not new; it was simply a reenactment of "the old divorce law."[5]

Owen's response side-stepped the fact that some of the 1852 legislation was indeed new—and extremely lenient. In particular, a minimal residency requirement became the linchpin of easy divorce in Indiana. According to the 1852 law, when petitioners filed for divorce, they simply had currently to reside in the county of filing. In addition, a petitioner's own affidavit was sufficient proof of residency. These provisions constituted an unusually loose residency requirement.[6]

The 1852 bill included another provision that contributed to ease of divorce in Indiana. Notification of divorce proceedings could be "served through publication" rather than being personally delivered to defendants. Specifically, notice of a divorce action could be published for three weeks in a "weekly newspaper of general circulation" in the venue where the action was filed. This procedure had many precedents. Because of the difficulty of locating deserted spouses, several colonies and states had experimented with a similar provision.[7] Although this action, called "presumptive service" or "notice by publication," was a common provision, in Indiana it was one more factor creating unusually easy divorce for both Indiana residents and divorce-seekers from less indulgent jurisdictions.

In addition to lenient divorce statutes, Indiana offered a central location and good railroad transportation. Indianapolis soon emerged as the hub of the Indiana divorce haven, for the city's bustling population offered anonymity to divorce-seekers. In addition, many Indianapolis attorneys actively courted the divorce trade, most of the city's newspapers accommodated customers who wanted to publish notices of their divorce suits, and the city boasted reasonably good shopping and recreational facilities.

It soon became clear that easy Indiana divorces could also have severe repercussions. An especially troublesome situation could result when one spouse divorced the other without his or her knowledge by simply publishing a notice in a locally circulated Indiana newspaper. Un-notified defendants were at a definite disadvantage, for they could not appear in court to oppose the divorce or to negotiate alimony, division of property, and child custody. In addition, an un-notified spouse, divorced *in absentia,* had no recourse—according to the 1852 statute, Indiana divorce decrees were irrevocable. Right of appeal regarding a divorce and its provisions was non-existent.

The lack of appeal was tested in the McQuigg case of 1854. After McQuigg obtained a divorce in Marion County, his wife attempted to have the divorce nullified. Because the McQuiggs lived in New York, she argued that her husband had established residence in Indiana solely to obtain a divorce. She also argued that an overruling of their divorce would deter "that large class of discontented or lecherous pilgrims seeking the Mecca of divorce, who turn their faces towards Indiana" because they would see that Indiana divorces could be reversed. Swayed by her arguments, the court voided the McQuiggs' divorce. This came as a great shock to McQuigg, who had already remarried. He quickly appealed the decision to the Indiana Supreme Court, which overturned the lower court's decision. The irrevocability of Indiana divorces stood. Much to McQuigg's relief, his 1854 divorce also stood.[8]

As a result of such controversies, a growing number of Hoosiers opposed easy divorce within their state. Many objected to the short residency requirement because they believed it attracted divorce trade from outside the state. Other critics, including women's rights groups, feared that un-notified, and thus absent, spouses would fare badly regarding child custody, alimony payments, and property settlements. Clergy, moralists, pro-family advocates, Quakers, and some conservatives believed that the rising divorce rate was a sign of growing immorality and the decline of the American family.

Pressured from many sides, Indiana legislators considered the issue of divorce reform. In 1853, they discussed several proposed bills, but failed to take action on them. In 1857, Indiana Supreme Court justices and Governor Joseph A. Wright called for revision of divorce statutes to free Indiana from out-of-state divorce-seekers and to halt mounting complaints. Some newspaper editors also joined the campaign. An 1858 editorial in the *Indianapolis Daily Journal* bitterly complained that the city was "overrun by a flock of ill-used, and ill-using, petulant, libidinous, extravagant, ill-fitting husbands and wives as a sink is overrun with the foul water of the whole house."[9]

In 1859, the legislature raised the residency requirement to one year's residence in the state before filing for divorce. Proof of residence beyond the petitioner's affidavit was also required. A divorce decree was still irrevocable, but a defendant in "notice by publication" divorces could ask the court to reconsider alimony, child custody, and property decisions.[10]

According to contemporary observers, these revisions had little effect on Indiana's divorce rate. In 1867, the French writer Auguste Carlier characterized Indiana as a state "where the law sympathizes strongly with conjugal misfortunes." He wrote, "Courts of justice are literally crowded with applications for divorce,—whose authors, it is true, are very often citizens of other States." Carlier concluded that Indiana's divorce statute was an "unwise law."[11]

Two years later, the notorious Schliemann divorce case conformed that Indiana divorce proceedings continued to be permissive. Heinrich Schliemann, a financier and the archaeologist who later discovered the site of ancient Troy, decided to divorce his Russian wife. In April 1869, he traveled to Indianapolis, where he established residence in a hotel. He then engaged several attorneys to file his petition and publish a notice of proceedings in an Indianapolis newspaper. Because his wife lived in St. Petersburg, Russia, it seemed unlikely that she would learn of his suit in time to oppose it. To hurry matters along, however, Schliemann purchased a house and part-interest in a starch factory. He used these investments to convince the Marion County court that he intended to remain in Indianapolis, with hopes that the court would agree to issue the much-coveted decree in less than the required year. Schliemann received his divorce on July 11; one week later he left Indianapolis with three copies of the decree securely in hand.[12]

The Schliemann case assumed scandalous proportions when Catherine Schliemann attempted to nullify the divorce. Her suit caught Schliemann in a difficult position; he had left his Indianapolis "home" and acquired a fiancée in Greece. Schliemann immediately took action to thwart Catherine's suit. Because his primary worry was his brief residency in Indiana, he sent complex instructions to his Indianapolis attorneys, directing them to validate his Indiana residency. He assured them that he and his future wife intended to live in Indianapolis. Catherine, however, ignored the issue of residency. She requested that the divorce hearings be reopened because of her absence from the trial. Schielmann was tremendously relieved to learn that according to the Indiana Code of 1852, Indiana divorce decrees were irrevocable. In addition, an 1859 revision allowed appeal in "notice by publication" cases only concerning matters of alimony, child custody, and property, all unmentioned in Catherine's appeal. Schliemann had provided well for his children and Catherine wanted nothing for herself. "Fortune," she wrote him in February, 1869, "is only agreeable when one has the free disposal of it and if one has not to submit for its sake to a limitation of one's personal liberty." Catherine's suit failed to reach the hearing stage, so the divorce stood.[13]

As a consequence of the Schliemann and other cases, a growing number of people inside and outside Indiana demanded enforcement of existing rules and enactment of restrictive provisions. In 1871, Governor Conrad Baker pleaded for changes in divorce statutes, especially because Indiana laws helped people evade the laws of their own states. He criticized attorneys who advertised for divorce-seeking clients in "the Atlantic cities" and called out-of-state petitioners "refugees and fugitives." In 1873, the legislature responded by eliminating the omnibus clause, adopting a provision requiring two years' residency in the state before filing for divorce, and allowing certain cases to be reopened within two years of the decree.[14] These changes constituted a serious move to end easy divorce in Indiana.

Was Indiana a freewheeling divorce mill that granted huge numbers of migratory divorces? Anecdotal evidence says yes. An 1858 letter from a District Recorder of Indiana to a New Yorker is one indictment of Indiana divorce laws. When the New Yorker inquired whether his wife had filed for divorce, the Indiana official replied: "There has not yet been an application for divorce made to our court . . . but I think we have divorced half of the citizens of your State, so that if we continue in the same train, I imagine, in a few years, we shall exhaust the marriages of New York and Massachusetts." In the same year, an editorial in the *Indianapolis Daily Journal* commented that every railroad town in the state was "full of divorce hunting men and women." Another journalist estimated that of the seventy-two divorce actions pending before the Marion County court fifty had been filed by out-of-state petitioners.[15]

William Dean Howells's novel *A Modern Instance* reinforced the view that Indiana courts carelessly granted easy divorces to out-of-state applicants. Howells described the plight of a wife denied the opportunity to represent herself in court because notice of divorce proceedings had been published in an Indiana newspaper rather than delivered to her directly. When she learned of the divorce, she exclaimed, "Oh, it's a cruel, cruel law!" Despite a dramatic dash to

Indiana, in 1879, she arrived too late to stop the divorce action. She learned to her dismay that the divorce decree was irrevocable.[16]

Other evidence regarding divorce in Indiana is less conclusive. Divorce statistics were collected beginning with the year 1867, but they defy concrete interpretation. For instance, one analyst speculated that although Indiana divorced more couples who had been married outside its boundaries than any other state, this might indicate a high rate of immigration into Indiana rather than a high rate of migratory divorces. Divergent opinions of Indiana divorce statistics make it difficult to either brand the state a divorce mill or absolve it of guilt.[17]

Whether or not Indiana was in reality a divorce mill, public concern about the Indiana situation was widespread. Although the number of migratory divorces granted by Indiana courts was probably lower than thought at the time, many people *believed* that migratory divorce-seekers escaped restrictive laws in their own states by obtaining divorces in Indiana. The specter of migratory divorce thus exacerbated the already emotional, volatile nature of the divorce debate in mid-nineteenth-century America, and inflamed those who believed that divorce should be cut back.

But was divorce the problem, or was it marriage? Utopian reformers argued that marriage needed revision rather than divorce. As early as 1826, Frances Wright's Tennessee commune, Nashoba, had rejected the concept of marriage. Wright and her followers believed couples should be bound by love and respect rather than legal ties. If love and respect evaporated, partners could separate. In this system, divorce was unnecessary to settle a couple's affairs and mark the end of their relationship.[18]

Other utopian communities also experimented with modified forms of marriage and family: New Harmony in Indiana, Fourierest phalanxes in Massachusetts, Icarian communities in such states as Texas and Illinois, and the Society of True Inspirationists in Amana, Iowa. But the most well-known experiments were probably the Shakers and the Oneida Community, both located in upstate New York.[19]

Under the leadership of Ann Lee, the first Shakers established a socialistic Christian community at Watervliet, New York, during the 1770s. The Shakers believed in total celibacy, a principle that struck most people as drastic and unworkable.

When Auguste Carlier toured the United States during the 1860s, he was intrigued by the Shakers: "It is worth noticing how all sorts of eccentricities rendezvous in this land of independence."[20] Eccentric they may have been, but their belief that celibacy would hasten the coming of the millennium attracted many converts during the early nineteenth century and Shaker communities multiplied.

Serious problems arose, however, when a married couple disagreed whether to join the Shakers—when one spouse wanted to join, but the other did not. The Shakers were afraid that some dissatisfied spouses would apply for membership only to escape their mates. Because Shakers wanted to avoid becoming a haven for such people, Shaker membership rules stated, "No believing husband or wife is allowed to separate from an unbelieving wife or husband, except legally, or by mutual agreement." Accordingly, Shaker officials frequently rejected the applications of married people. One of these, a woman with four children, was turned away until she was free "of bondage to that drunken man." Moreover, Shaker authorities flatly refused to accept a man or a woman who had "abandoned his or her partner, without just and lawful cause."[21]

Still, some spouses succeeded in joining Shaker societies although their mates refused to do so. Consequently, in states where Shaker communities were located, legislators enacted special statutes recognizing as a ground for divorce the departure of a spouse for a celibate religious society. Kentucky passed the first "Shaker" law in 1812; Maine followed in 1830, New Hampshire in 1842, and Massachusetts in 1850. In Connecticut, joining a celibate sect fell into the omnibus clause passed in 1849.[22] Inadvertently, then, Shakers expanded the list of grounds for divorce in several states and helped escalate the growing controversy regarding divorce in the United States.

The Oneida Community advocated a very different type of relationship between its female and male members. Oneida leader John Humphrey Noyes had begun to develop his unconventional beliefs in 1834. In 1849, he founded the Oneida Community. Here, Noyes established complex marriage, meaning that every woman in the community was married to every man, and every man to every women. Noyes hoped to eliminate competition, jealousy, and inequality between women and men so that spouses would no longer feel "ownership" of their partners. In complex marriage, the need for divorce would be eradicated.[23]

In 1850, an Oneida tract equated marriage with slavery and argued that marriage was "contrary to natural liberty." Marriage, "a cruel and oppressive method of uniting the sexes," led to abuses of women who were the weaker parties in marriages. These violations of women's rights and freedoms were similar to the "cruel lot" of slaves. To Oneida theoreticians, marriage was little more than "a huge Bastille of spiritual tyranny where men and women have the power to debar each other from their rights of conscience, and the enjoyment of their religious faith."[24]

Members of the Oneida Community continued to practice and preach complex marriage for many years, but they eventually abandoned their bold experiment. After Noyes retired in 1877, the community adopted monogamous marriage; in 1880, it disbanded.[25]

## South Carolina: The Other End of the Spectrum

South Carolina stood alone at mid-century; its leaders refused to experiment with divorce. Because they believed that families were the bedrock of social harmony, they were determined to keep families intact. They would have agreed with the South Carolina essayist who maintained that family order led to "virtue in the state," while divorce created some of the "greatest mischiefs" plaguing modern societies.[26]

Their thinking proved to be fallacious, however, for prohibiting divorce failed to maintain family order and social virtue in South Carolina. Marriages crumbled and couples parted despite the prohibition. The situation in South Carolina seemed to prove some reformers' point that the problem demanding attention was marriage rather than divorce, for even when divorce was non-existent, marriages broke down.

Several especially distraught spouses pleaded their cases before the South Carolina General Assembly even though it had no power to dissolve their marriages. In 1830, Curtis Winget applied for a divorce on the ground of his wife's adultery with men named Mitchell and Bailis. At one point, Winget's wife eloped with Mitchell, then returned to Winget. She then bore a child that she claimed Bailis had fathered. Members of the assembly agreed that Curtis Winget deserved a divorce, but they lacked the power to grant it. Subsequently, four other South Carolinians who entered divorce pleas met with the same fate.[27] These five marriages stayed intact legally, but contributed little to the order and harmony of the state of South Carolina.

Other disgruntled mates took their cases to chancery courts, which sometimes granted annulments. Chancellors also decreed divorces of bed and board, which provided alimony and child support, but restricted the parties from remarrying.

Yet other spouses ended unsatisfactory marriages by deserting their mates. After Ann Sims's printer husband left her for an unknown western location, she was required to wait seven years for legal resolution of the situation. After seven years, a desertion could become a divorce of sorts if authorities declared the deserting spouse legally dead and granted the deserted mate permission to remarry. But tangled family webs often resulted from this rule. One example was that of a deserted Lexington County wife who waited seven years, then married a man named Kennerly. When Kennerly died, his heirs maintained that her former husband was still alive and contested her right to Kennerly's property. Undismayed, she claimed that Kennerly was actually her third husband. She explained that her first husband had abandoned her. She then married a second husband, but this marriage was invalid because her first husband was still alive. Her first husband died shortly before she married Kennerly, making the third marriage valid. By the time she convinced the court of the truth of her complicated marital relationships and her right to part of Kennerly's estate, she had married a fourth husband.[28]

Still other unhappily married South Carolinians sought out-of-state divorces, a practice that created legal problems if they returned to South Carolina. Did South Carolina courts have to recognize out-of-state divorces? What was the legal status of children produced by a South Carolinian who had remarried after an out-of-state divorce? And, did first wives have sole claim on the estate of a man who had divorced out-of-state and remarried in South Carolina?

According to the Constitution, each state is required to give "full faith and credit" to the acts of another state. At the same time, however, the right of each state to regulate the marital status of its citizens implies that states can refuse to recognize out-of-state divorces if they chose to do so. In an 1852 case involving a South Carolinian who divorced in Georgia and returned to South Carolina, the court refused to recognize the Georgia divorce. Instead, it restated the principle that a South Carolina marriage was "indissoluble by any means."[29] This decision meant that children of subsequent marriages would be considered illegitimate, and first wives would have sole claim on the estates of remarried men.

Still other South Carolinians accepted the ban on divorce and stayed within their marriages. Frequently, however, they dealt with their plight by establishing adulterous liaisons. This raised legal questions about what was to happen to the paramour of a married man after his death: Would she have inheritance rights to his estate? Could he leave his entire estate to her and exclude his wife?

South Carolina law protected both wives and mistresses by stipulating that a man must leave a portion of his estate to his wife and could leave no more than one-quarter to his mistress. In 1867, Auguste Carlier scathingly remarked that South Carolina law had legalized "concubinage." Adultery was not prohibited, punished, or recognized as a cause for divorce. Instead, it received "a sort of sanction" from legislators who devised a "special statute" specifying "what portion of his property a married man may give to his concubine." But, because male legislators neglected to create a comparable law for women and their paramours, estates of adulterous wives went to their husbands, with no share to their lovers.[30]

In 1868, South Carolina policy-makers initiated a brief experiment with divorce. The state's 1868 constitution included a divorce clause. Two years later, an 1870 statute specified adultery and desertion as grounds for divorce, and permitted courts to grant alimony and child support. But on December 20, 1878, legislators repealed these statutes. Not one divorce had been granted during the ten years that the possibility existed in South Carolina.[31]

South Carolina's prohibition of divorce primarily affected white men and women. Because African Americans were denied access to white law and marriage, they created their own mechanisms. African American churches sometimes granted divorces to discontented slaves and to those whose mates had been sold. African American communities also established rules that allowed dissatisfied couples to separate. And some planters regulated marriage and divorce among their slaves. In Edgefield, for example, planter James Henry Hammond permitted divorce and held regular divorce hearings: "Had a trial of Divorce and Adultry cases . . . Separated Moses and Amy finally."[32]

Most white South Carolinians would probably have argued that their system, which prohibited divorce, was far superior to the African-American system, which allowed divorce. Yet white attempts to maintain family order by banning divorce failed to achieve its ends. Distressed husbands and wives circumvented the no-divorce rule through desertion, informal separation, divorce of bed and board, and migratory divorce; these were disruptive and sometimes illegal solutions. Rather than maintaining order, the lack of divorce in South Carolina created disorder.

## The Divorce Debate Escalates

As divorce became more widespread, it received far more public attention than ever before. In general, people supported one of two major positions.

One faction argued that American society must return to the traditional view of marriage and divorce. It advocated religious tenets that dictated the sacramental, lifetime nature of marriage and eschewed divorce except in extreme cases. To members of this group, *divorce* was a social problem. In their view, the larger good of society—the need to maintain families, harmony, and stability—was more important than marital dissatisfaction of individual citizens.

A second group argued that divorce reform meant easing and widening divorce laws. To them, *marriage* was a social problem. Because of their concern for the rights of individual Americans, they saw divorce as a liberating device for those in distressing marriages. They believed that the welfare and happiness of individuals must come before the good of society. And they argued that a mate persecuted by a wife's or husband's abuse, alcoholism, or other harmful behavior should have the right to go free.

A great debate ensued between the anti-divorce and pro-divorce factions. Although the debate failed to reconcile these opposing positions, it articulated and clarified key issues. Was marriage a lifetime union or a dissolvable contract? Should divorce be prohibited entirely, be allowed only on the biblical grounds of adultery, or be easily available?

Joel Prentiss Bishop's two-volume analysis of divorce in the United States was an early entry in this debate. Bishop's *Commentaries on the Law of Marriage and Divorce* (1852) criticized legislative divorce and slack procedures but did not oppose divorce itself. Nor did Bishop believe that divorce threatened the future of the family and American society. Bishop used the example of Connecticut, a state that liberally granted divorces. He maintained that in spite of the "liberty of divorce," or perhaps because of it, Connecticut society was marked by "domestic felicity and purity, unblemished morals, and matrimonial concord and virtue." He explained that people who believed that marriage was "a religious sacrament, and indissoluble" could preserve their principles by refusing to divorce. If their mates divorced them, they were "still permitted to retain the seal of the sacrament pure and undefiled in their consciences" by refusing to remarry.[33]

Bishop's comprehensive treatment of divorce was only one contribution to the escalating nationwide discussion. Magazine articles presented arguments for and against. Protestant and Catholic tracts and statements warned against divorce. And in 1859, what was probably the first divorce self-help book, *How to Get a Divorce*, advised divorce-seekers concerning divorce laws.[34]

A series of articles and letters in the *New York Tribune* constituted the longest running segment of the debate. The first round occurred between November 1852 and February 1853, when the newspaper's editor, Horace Greeley, debated writer and lecturer Henry James, Sr., and free love advocate Stephen Pearl Andrews. James supported divorce, arguing that "manifest public welfare" demanded its availability. He believed that spouses would be more productive if liberated from debilitating marriages. This view was far too conservative for Andrews, who asserted that governments should stay out of marriage and divorce entirely. If no "legal bondage" tied people to each other, divorce would be unnecessary. Thoroughly appalled by James's and Andrews's contentions, Greeley argued for "indissoluble marriage" except in cases of adultery. To him, "the highest good of the community" demanded "the subjection of individual desire and gratification."[35]

Greeley rekindled the *Tribune* debate in March 1860, when he blamed Robert Dale Owen and his "lax principles" for helping create a divorce mill in Indiana. A spirited interchange ensued. Greeley reiterated the biblical view of marriage as a lifelong contract terminable only in cases of adultery. To him, marriage was "a solemn engagement to live together in faith and love till death." Owen countered that the biblical view of marriage and divorce had been devised for ancient Judea rather than present-day America. In the modern world of the 1860s, Owen insisted, divorce would purify marriage and help it become "the holiest of earthly institutions." When the battle dragged to a halt some two months later, neither side had budged a whit.[36]

In the meantime, women's rights advocate Elizabeth Cady Stanton declared herself a supporter of divorce. Although she had occasionally spoken on behalf of divorce before 1860, in that year she presented her first comprehensive public statement regarding divorce to a women's rights convention in New York City. Stanton told her audience that "an unfortunate or ill-assorted marriage is ever a calamity, but not ever, perhaps never, a crime." She then maintained "when society or government, by its laws or customs, compels its continuance, always to the grief of one of the parties, and the actual loss and damage of both, it usurps an authority never delegated to man, nor exercised by God himself." Ridiculing the idea of indissolvable marriage, Stanton sharply criticized Greeley: "I know Horace Greeley has been most eloquent, for weeks past, on the holy sacrament of ill-assorted marriages, but let us hope that all wisdom does not live, and will not die with Horace Greeley."[37]

Stanton was upset because men made and administered laws concerning divorce. Men, she charged, "have spoken in Scripture" and "in law." As husbands, men decided the "time and cause for putting away" their wives. As judges and legislators, they exercised "entire control" over divorce proceedings. She preferred that women, "the mothers of the race," participate in making such decisions.[38]

Stanton set aside her causes during the Civil War, but after the war she resumed her pro-divorce campaign. In articles and speeches, Stanton condemned patriarchal marriage, calling instead for equality between wives and husbands. She also damned the "slavery" of a bad marriage and stressed the right of individual Americans to seek liberty from such a marriage. In the pages of *The Revolution*, a women's rights' journal that she edited, Stanton defended the right of wives and husbands to flee marriage through divorce. Each "slave" who fled a "discordant marriage" was a source of joy to her. She predicted, "With the education and elevation of women we shall have a mighty sundering of the unholy ties that hold men and women together who loathe and despise each other."[39]

Other women's rights leaders diverged widely in their stands on divorce. Congregational minister Antoinette Brown opposed divorce because the Bible defined marriage as permanent and indissoluble. Jewish immigrant and reformer Ernestine Rose, however, cited mismatched spouses, physical and verbal abuse, drunkenness, and other manifestations of marital disruption as good reasons for allowing divorce. In her view, divorce provided an escape route for women from bad marriages—from "legalized prostitution." Like Stanton, Rose argued that people had a right to seek happiness even if that meant divorcing one spouse and marrying another.[40]

The issue of free love was often tied to divorce, but Rose stopped short of advocating free love. To her, despite its problems and occasional breakdowns, marriage was still superior to the "crimes and immoralities" of free love. But radical reformer Victoria Woodhull advocated both divorce and free love. In *Origin, Tendencies and Principles of Government,* she envisioned a perfect society. The marriage tie, which was degrading to women, was to be replaced by free love, and divorce would be unnecessary.[41]

Other women's rights leaders advocated divorce because they thought it offered salvation to wives of alcoholics. It was widely believed that alcoholic husbands fell into drunken rages and mercilessly beat their wives and children. "A wife he thrashes, children lashes," was an oft-repeated bit of doggerel. Consequently, letters demanding easy divorce for wives of alcoholics regularly appeared in the pages of Amelia Bloomer's temperance journal, *The Lily.* In a letter to *The Lily,* Jane Swisshelm, a women's rights advocate and disillusioned wife, recommended that abusive, drunken husbands be whipped and that women who remained married to alcoholics be committed to insane asylums. Gradually, lawmakers responded to such protests. By 1871, thirty-four states, two-thirds in the South and West, permitted divorce for drunkenness.[42]

As the 1860s waned, the divorce debate raged fiercely. Yet when the McFarland-Richardson divorce case erupted in 1867, it stunned the nation, escalated the divorce debate to a new high, and forced even the most committed advocates of divorce to admit that divorce was often fraught with peril.

The McFarland-Richardson case involved an especially messy series of events climaxed by a murder and a deathbed marriage. Abby Sage McFarland and Daniel McFarland, spouses and parents of two sons, were its principals. Because Daniel consumed far too much alcohol, he was ineffective as a lawyer, land speculator, and in other jobs. To supplement the family's inadequate income, Abby gave dramatic readings and acted on the stage in bit parts.

As she grew more alienated from Daniel, Abby formed a relationship with *New York Tribune* reporter Albert D. Richardson, who lived in the same boardinghouse as the McFarlands. When Abby left Daniel early in 1867, she and Richardson decided to marry. Abby went to Indiana, where she obtained a divorce after sixteen months' residency. When she returned to New York, an outraged Daniel accosted Richardson and shot him in the stomach. Richardson died a week later, on December 2, 1869, only two days after he married Abby. In a final twist to the story, divorce opponent Horace Greeley cared for his dying friend Richardson and arranged for renowned minister Henry Ward Beecher to wed Richardson and Abby McFarland.[43]

The McFarland-Richardson affair had all the elements of a juicy scandal: sex, intrigue, and blood. In addition, the great moralists, Greeley and Beecher, opened themselves to attack and controversy. A *New York Times* editorial disparaged Greeley and Beecher for their "open scorn of the marriage tie and the total disregard of all principles of justice." Members of the National Woman Suffrage Association attacked Greeley and his newspaper for its "pernicious influence on the divorce question and the civil rights of women." Another contingent of suffragists, however, praised Greeley for supporting his dying colleague. A few days later, Elizabeth Cady Stanton asserted that a woman, in this case Abby McFarland, "has a right to choose between a base, petty tyrant and a noble, magnanimous man."[44]

Public indignation reached a fever pitch when Daniel McFarland was acquitted of murder early in 1870. McFarland's attorney invoked the Bible in support of a husband's right to fend off interference with his wife, a defense that elicited hisses from women in the audience. After being freed on a plea of temporary insanity, McFarland received custody of his son.[45]

Stunned and outraged at what she saw as the unfairness of New York law, Stanton protested the court's decision in an address to two thousand women in New York City's Apollo Hall on May 17, 1870. Stanton stated that she felt compelled to oppose the "unjust decision in our courts" and the "scurrility of the press." She also attacked the "popular idea of marriage." In a little-known speech delivered to a small gathering of women and men later that year, she argued for free love rather than marriage. In her eyes, free lovers were "among the most virtuous of women and men" because they relied upon affection rather than legal entanglement to bind them together. She hoped that people who were "dabbling" with woman suffrage understood that the next step after equality was freedom: "in a word Free Love." And she warned her audience that should they "wish to get out of the boat," they should "for safety sake get out now, for delays are dangerous."[46]

Throughout these tempestuous weeks, Stanton received support and encouragement from her long-time friend Theodore Tilton. As editor of *The Independent,* a religious journal, Tilton condemned New York divorce statutes which, by allowing only adultery as a cause of divorce, tied wives to alcoholic husbands who made their lives "constant agony." He urged New York legislators to enact lenient divorce statutes like those in New England rather than showing "a blind allegiance to a superstition which never had any other foundation than a long-exploded interpretation of a Scriptural text."[47] Tilton was unaware that he would soon be embroiled in one of the most sensationalized adultery trials in American history.

It was Victoria Woodhull—stockbroker, editor, women's rights supporter, and free love advocate—who first broke the story of Tilton's wife's affair with minister Henry Ward Beecher, in the November 2, 1872, issue of her newspaper, *Woodhull's and Claflin's Weekly.* Woodhull was disgusted by what she saw as Beecher's hypocritical, moralistic stance in the pulpit of Plymouth Church in Brooklyn, New York. After Woodhull's lurid report of Beecher's liaison with Elizabeth Tilton, Beecher's congregation stood behind him. In 1874, when Theodore Tilton sued Beecher for alienation of affections, the jury voted for Beecher.

The following year, the Tiltons divorced. Theodore gave up the vestiges of his career and moved to France, while Elizabeth became a bitter, lonely recluse. In the meantime, Woodhull took to the lecture circuit to offer the American public additional details about the Beecher-Tilton affair. Perhaps as a result of the stress that Woodhull's attack on Beecher created in her life, Woodhull's second marriage ended on September 18, 1876. After charging her second husband, James H. Blood, with adultery and seeing him barred from remarrying under New York divorce law, Woodhull left the United States and relocated in England.[48]

Opponents and proponents of divorce drew vastly different conclusions from these titillating events. Opponents of divorce, including followers of Horace Greeley, who died shortly after Woodhull exposed the Tilton affair in 1872, blamed existing divorce laws, women's rights, and free love for breakdown in marriage in particular and society in general. Supporters of divorce, however, suggested that existing divorce laws needed revision and expansion. Stanton, who had known the Tiltons' secret before it became public, urged New York lawmakers to add other grounds for divorce to the sole cause of adultery. Compelling "unhappy husbands and wives" to live together, "produces, ever and anon, just such social earthquakes as the one through which we are now passing."[49]

Rather than supporting Stanton, many suffragists took an opposing view. In their meetings and conferences, rank-and-file members of women's rights groups passed resolutions reaffirming customary conceptions of marriage. They also repudiated easy divorce and free love. Distressed that women's rights groups had been tarred with the brush of free love, they wanted to set the record straight on their position. In 1871, the Henry County Woman Suffrage Society of Iowa passed a ringing declaration on the side of tradition: "In presenting the claims for women we recognize the doctrine and sentiment in regard to marriage and all other subjects as set forth in the teachings of Jesus Christ." A few months later, the Polk County Society, also in Iowa, resoundingly proclaimed: "As Christian women . . . we do not believe that liberty means license, and that we have no sympathy with any sentiments taught by any of the champions of woman suffrage which have a tendency to loosen the bonds of Christian people which we believe to actuate the vast majority of the women of America."[50]

Suffrage leaders also believed that their efforts had been tarnished and diffused in far-off, but highly visible, New York City. Annie Savery, a women's rights leader in Iowa, protested, "The woman suffrage party of Iowa is made up of the mothers, wives, and daughters who believe that the marriage bond is to the social what the Constitution is to the political Union."[51]

In Illinois, also in 1871, editor and reformer Jane Swisshelm helped sever the Illinois Woman Suffrage Association into two factions over the issues of divorce and free love. Shortly after delivering a blistering anti-free love speech to the association, she accepted the presidency of a splinter group, the religiously oriented Illinois Christian Woman Suffrage Association. Swisshelm then spearheaded the Christian Woman Suffrage Association's campaign against divorce and free love. Paradoxically, Swisshelm had escaped a burdensome marriage and developed her career as a journalist by divorcing her husband.[52]

Were many minds changed by this debating and haranguing? Probably not, but it is clear that divorce gained a taint of disgrace and stigma during these years. Despite this development, however, more people than ever sought divorces. Even as the anti-divorce faction campaigned against it, divorce increased. If judged by the upward climb of the divorce rate in the United States at mid-century, the pro-divorce faction had gained an edge.

Unfortunately, many of these divorce-seekers fared poorly because they were governed by laws and policies forged out of controversy and compromise rather than concern for their best interests. The divorce debate had diverted attention and energy from the creation of effective, workable divorce laws, procedures, and policies. Anti-divorce and pro-divorce factions had argued over the wisdom of divorce, but they seldom mentioned the deleterious effects of current laws upon divorce-seekers. Such issues as alimony and child custody demanded immediate attention, but the divorce debate focused on divorce itself, leaving these issues to develop in an inconsistent, unplanned fashion.

## The Growth of Divorce

Despite the millions of words written and spoken against it, the stigma attached to it, and the tears shed during it, divorce continued to spread in mid-nineteenth-century America. People racked by the pain of a distressing marriage sought divorces even if they, their families, and their ministers believed it an immoral solution. If necessary, some of them relocated in lenient states that would grant them divorces and give them an opportunity to get their lives back on track.

The Civil War especially boosted the divorce rate. By 1861, bitter arguments concerning sectional power and slavery split the nation. For the next four years, the Union and the Confederacy struggled to decide if they would dissolve the marriage between them. The decision in the Civil War was for indissolvable union, for a marriage of states that would endure through better or worse.

Yet, as the Union and Confederacy decided to remain wed, more American couples decided to divorce. In all likelihood, the divorce rate rose during the war years because many people married in haste. One commentator speculated that the Civil War created an increase in "ill-assorted" marriages that were little more than hurried alliances between attractive young women and men romantically clad in military garb.[53]

The exigencies of war also placed great stress upon marriages. Thousands of husbands and wives went to the front as soldiers, nurses, and sanitary agents distributing food, medicine, and clothing. Thousands more took jobs in factories and government offices. Those who remained at home tried to keep farms and family businesses afloat, and their families intact. In the face of such factors as new experiences and loneliness, some people remained faithful to their mates, but others did not do so.

As a result of such wartime stresses, the divorce rate jumped noticeably. In 1860, 1.2 of 1000 marriages ended in divorce. In 1864, 1.4 ended in divorce; in 1866, 1.8 did so.[54] In 1867, Auguste Carlier remarked on Americans' seeming penchant for divorce: "the number of divorces is very considerable." He appeared pleased that women obtained more divorces then men. "To the honor of American women it must be said," he wrote, "that the majority of the divorces are granted at their request, and not against them." According to Carlier, benevolent divorce laws restored people's personal liberty and allowed "the parties to seek another union, better assorted."[55]

Twenty years later, the U. S. Bureau of Commerce and Labor collected statistics that proved the accuracy of Carlier's observations. On March 3, 1887, members of Congress directed Commissioner of Labor Carrol D. Wright to collect marriage and divorce statistics in all states and territories for the years 1867–86. The Wright report, a sizable two-volume compendium, disclosed the following increase in divorces during the Civil War period:

| Years | Divorces Granted | % Increase Over Previous Period |
|---|---|---|
| 1867–71 | 53,574 | |
| 1872–76 | 68,547 | 27.9 |
| 1877–81 | 89,284 | 30.3 |

These statistics indicated the number of divorces had increased faster than the population and faster than the married population. "The most significant tendency," Wright pointed out, "is the marked persistency of the increase in the divorce rate."[56]

During these years, women obtained approximately two-thirds of all divorces granted in the United States and men obtained one-third. Wright speculated, "The wife has a legal ground for divorce more frequently than the husband" because "certain well-known and comparatively common grounds are more readily applicable against the husband than against the wife." These grounds were non-support and cruelty; women obtained five out of every six divorces granted for these causes.[57]

Commissioner Wright was distressed that race of divorce litigants had to be omitted from his report. Although he had instructed field agents to record race, they soon discovered that it was rarely noted in divorce records. Wright tried to determine white and black rates of divorce by analyzing predominantly white and black counties. After discovering that black counties had a lower rate of divorce than white counties between 1867 and 1881, he concluded, "It is hard to believe that divorces among negroes could have attained much importance at that early date."[58] It seems likely that African Americans probably continued to obtain divorces from their churches as they had done for years rather than avail themselves of white law so soon after emancipation.

The Wright report was also sensitive to other variables affecting the divorce rate. It showed that western states granted the highest number of divorces, while northern states were next and southern states were last. Because Wright was interested in divorce in urban and rural areas, he was disappointed that divorce records failed to specify whether divorcing couples lived in cities or in the countryside. By analyzing counties containing cities, he determined that the divorce rate was higher in large cities than in small cities and rural districts. The report touched only lightly upon social class by indicating that divorcing husbands held a wide variety of jobs, and it neglected to discuss ethnicity.[59]

The Wright report made it very clear, however, that the divorce rate in the United States far outdistanced those in other nations. England, for example, had only begun granting judicial divorces after Parliament passed the Divorce Bill of 1857. By 1886, English courts were granting slightly over 400 divorces per year, a considerably smaller number than American courts.[60] Unarguably, divorce was a more widespread phenomenon in the United States than in any other country in the world.

Despite the growth of divorce and the furor concerning it, most Americans continued to see marriage—accompanied by love—as an important part of their lives. Clearly, love had a firm grip on the minds and hearts of mid-nineteenth-century Americans. One young Civil War soldier, writing home about his admiration of "fair heroines," said that after the war ended he intended to find a "true and intelligent southern lady" to share his remaining days "on this gloomy and uncertain sphere." He was convinced that love would erase his memories of death and destruction as well as obliterating his storehouse of sadness and pain. Some years later, a California woman similarly lauded the efficacy of love: "Truly I do believe that love is all in life."[61]

Certainly, most courting couples believed that love was a crucial ingredient of marriage. They expected love to bridge the gap between the separate worlds of men and women, to create a closeness between mates that would negate the dissimilarities in men's and women's values, attitudes, and activities. The presence of love would reconcile a man's ties to the world outside the home with a woman's ties to home and family. Ultimately, love would prevent spouses' differences from creating discord in their marriages.[62]

Paradoxically, it was increasingly apparent that love was incapable of solving all problems and that nuptial bliss was often elusive. Tracts and essays exposed women's growing disillusionment with wifehood, and some writers even challenged the sacred nature of motherhood. One flatly stated that wife- and motherhood were less than rewarding jobs; anyone could do them. Reformer Catharine Beecher suggested that women should be trained for these domestic roles, for more job satisfaction would surely occur if wife- and motherhood were professionalized.[63]

A few novelists also disclosed Americans' growing dissatisfaction with marriage. T. S. Arthur warned that wives and husbands frequently failed to get what they expected in mates. According to Arthur, wives sought a number of attributes in husbands, including faithfulness, courage, fortitude, strength, and willingness to confide in a wife. Husbands expected women to be "true" wives who sustained them in time of stress and avoided criticizing them. When such expectations went unmet, a couple's wedding day might be followed by the rapid and unexpected onset of "sober realities." Husbands did not always communicate as openly as women wished them to; many became overbearing and bossy once the marital knot was tied. Wives could be jealous, petty, and petulant rather than self-sacrificing. Such traits, Arthur warned, could put more pressure on a marriage than a husband and wife were willing or able to handle.[64]

The rising divorce rate demonstrated that growing numbers of people refused to live with the eccentricities of their mates' personalities. Instead of erasing their difficulties, love often evaporated, leaving them to confront their problems without its assistance. When disenchanted spouses sought divorces, they increasingly charged their mates with broadly defined marital errors.

In particular, the charge of cruelty was rapidly becoming the ground of choice in mid-nineteenth-century America. New Hampshire had adopted the cruelty plea in 1791, Vermont and Rhode Island in 1798, Ohio in 1804, Kentucky in 1809, Pennsylvania in 1815, Delaware and Michigan in 1832, Iowa in 1839, Texas in 1841, and Kansas in 1855. By 1886, only six states refused to accept cruelty as a ground for divorce.[65]

By mid-century, Americans were becoming aware of the existence of wife abuse and were speaking out against it. Women's groups, especially temperance organizations, campaigned and lobbied against wife abuse. Many men took a stand against it as well. In 1862, the Commissioner of Agriculture speculated that the mental instability thought to be prevalent among farm wives had more to do with their husbands' abusive treatment of them than it did with isolation, unstable finances, and crop losses. Some years later, a male newspaper editor reported an instance of wife-beating and labeled the husband a "worthless whelp."[66]

Sexual submission to husbands was also challenged, especially by female divorce-seekers. Because there were no laws against marital rape, female petitioners who believed that they had been sexually abused used the ground of cruelty. They explained that their husbands had endangered their health and well-being by demanding excessively frequent sexual relations. In such cases, judges became moral arbiters who had to decide how much sex was too much, whether a husband had a right to his wife's body at all times, and whether a husband could force his wife to have sexual relations against her will.

Given Victorian ideas about protecting vulnerable women and controlling lustful men, it is little wonder that many male judges began to grant divorces to women who accused their husbands of sexual cruelty. But judges seldom accepted husbands' charges that their wives refused to engage in sexual relations as a ground for divorce. Consequently, male divorce petitioners who accused their wives of cruelty in refusing to engage in conjugal relations had a notably lower success rate than did female petitioners who charged their husbands with excessive sexual demands.[67]

The ground of cruelty also began to encompass verbal and mental abuse. Many judges agreed that a husband who hurled such epithets as "whore" and "slut" at his wife was committing cruelty. And when a husband falsely accused his wife of adultery, he was damaging her health, well-being, and reputation. Some judges even defined a spouse's refusal to communicate openly as a form of cruelty. Harsh words, coldness, an unwillingness to communicate, and other thoughtless acts were considered cruel treatment. In 1849, for example, Pennsylvania Judge Edward King stated that physical violence was only one type of destructive behavior. To him, "a course of humiliating insults and annoyances, practised in the various forms which ingenious malice could readily devise" could be extremely harmful "although no physical violence ensued." Justice John M. Scott of the Illinois Supreme Court concurred: "Cruel treatment does not always consist of actual violence."[68]

The increased use of the ground of cruelty lends itself to a number of interpretations. Undoubtedly, some petitioners used this charge to avoid the personal revelations harsher grounds required, especially adultery. Increased use of the cruelty plea also suggests that people were beginning to object to physical and verbal mistreatment in marriage and wanted the right to leave an abusive spouse rather than endure his or her behavior. Their expectations of marriage appeared to be higher than those of their parents and grandparents.

Still, it was more than rising expectations that pushed these people into divorce court; it was often a serious and sometimes life-threatening situation that did so. During the mid-nineteenth century, the decision to seek a divorce was still a momentous one, especially for women.

In particular, because women were usually economically dependent on men, most were concerned about their economic survival after divorce. Thus, the rising number of divorces brought the problem of alimony into the limelight. Also, divorced women increasingly made up a sizable group of people impoverished because alimony was awarded only erratically and, when awarded, frequently went unpaid.

Several women novelists took up the cause of divorced women's need for alimony. In 1876, E.D.E.N. Southworth castigated "cruel laws" that made woman an "idiot" and a "chattel" in the eyes of the law and forced her to beg for alimony. Southworth blamed men for casting divorced women into poverty. In her view, inadequate divorce laws, enacted by woman's "natural lover and protector, man," forced many divorced women into indigence. Novelist Mary Grace Halpine also condemned the contemporary system of alimony. In *The Divorced Wife*, Halpine painted a distressing picture of a woman who bartered her claim to alimony for custody of her children, and of another who was reduced to pennilessness by divorce.[69]

Child custody awards were another thorny issue. During mid-century, courts began to reject the longstanding idea that fathers "owned" children and should automatically receive custody of them. Instead, many judges adopted the "tender years" principle, meaning that mothers, except those who were totally unsuitable, would receive custody of young children. In 1860, a New Jersey statute expressed the tender years view of child custody: "The mother is entitled to the custody of her children under the age of seven unless it affirmatively appears that, in her custody, they would be exposed to either neglect, cruelty, or the acquisition of immoral habits and principles."[70]

The tender years axiom probably created more problems than it solved. For instance, it exacerbated the adversarial nature of divorce proceedings. If a mother was the guilty party, her chances of getting child custody were slimmer than if she was the innocent party. Some people felt this was unfair; that a mother should have custody of young children despite her marital errors. One woman declared, "it is surely time that the law that gives a man the custody of the child his wife has borne in pain, even when she gets a divorce for adultery, was wiped out & a decent one put in its place."[71]

The tender years rule led to other problems as well. Judges could split brothers and sisters by placing younger children with their mother and older children with their father. Some judges also tended to give mothers custody of girls and fathers custody of boys. Judges who gave a mother custody might deny her alimony, so she had to support herself and her children by taking whatever job she could find, falling back on her family, or accepting charity.[72]

Obviously, the process of divorce could be destructive for men, women, and children. Because of unclear and inadequate provisions, divorce laws often subverted divorce-seekers' quests for personal liberty and happiness. Although disillusioned spouses turned to divorce as a balm for their wounds, as a beneficent process that would adjudicate their tangled relationships and give them another chance to find the love they had lost, they frequently emerged emotionally bruised and financially burdened.

The argument that a marriage vow was a dissolvable contract gained ground by the late 1870s. It was aided by such factors as urbanization, industrialization, changing expectations of marriage, women's expanding roles, and wartime stress that had placed pressure on mid-nineteenth-century marriages—pressure that increasingly resulted in divorce.

In addition, change was the byword of the day: free lovers joined communes and eschewed marriage; women campaigned for an expansion of their rights; labor confronted their employers with new demands; and Transcendentalists and other intellectuals spoke of equality, serenity, and fulfillment. Into this world of change and challenge, of striving for betterment, of turning old ways on their heads, divorce fit remarkably well. Splitting had long been an acceptable way for Americans to deal with problems; now it was a progressive and forward-minded one as well.

But, at the same time, many Americans who feared the impact of divorce on American society and the family opposed its proliferation. With Greeley, they believed too many people were getting "unmarried nearly at pleasure." To them, divorce seemed like a scourge spreading across the nation. The remarkable growth of divorce in the American West and provocative tales about western divorce mills pushed these anti-divorce Americans to the brink; many of them would soon demand reform.

# 4

# *Divorce and Divorce Mills in the American West*

☐ "Just cause and good provocation" convinced Reuben Ward of Howard County, Missouri, that he and his wife should "tear" themselves "asunder." In a whimsical notice in the Missouri *Intelligencer,* June 19, 1824, he advised her to get a divorce: "when you readest this, suppress thy sobs, sue out a divorce, and set thy cap for another and a more happy swain, while I roam through the world sipping honey from the bitter or sweet flowers that chance may strew in my path."[1] With this lighthearted farewell, Ward joined thousands of others who left their homes and spouses behind them.

Ward's attitude was far from unusual in the nineteenth-century American West, a region widely known for its dedication to individualism, breaking ties, and reshaping institutions. Although settlers carried established ideas and institutions westward, most refused to be bound by them. Instead, they revised customary procedures whenever it suited their purposes. In addition, western settlers frequently acted in haste. Because they were anxious to establish government and other institutions, westerners often skipped time-consuming deliberations. Haste set the state for the adoption of permissive divorce statutes and short residency requirements on more than one occasion.

Western states and territories soon gained notoriety for their broadminded or, as some said, decadent divorce laws. The West was widely known for divorce laws that were, according to an 1867 observer, "very liberal; seldom compelling men or women to remain in marriage bonds which they wish severed."[2]

The American West seemed to provide a hothouse environment for the institution of divorce. Here, the divorce rate rose faster than in northeastern or southern states. Even if migratory divorces were subtracted from total divorces, the West's divorce rate considerably exceeded those of the Northeast and South. In addition, one town after another gained a national, and sometimes international, reputation as a divorce mill during the latter part of the nineteenth century. Gradually, lenient western divorce laws and colorful western divorce mills helped convince many Americans that the time had come to regularize and control divorce in the entire United States.

### Divorce and Desertion

In 1908, Commissioner of Labor Carroll D. Wright reported that "the divorce rate increases as one goes westward." Wright and the staff of the Census Bureau collected statistics beginning with the year 1867 that revealed the ratio of divorces to population increased faster in western states than in any other region of the United States.

The growth of divorce in the West is even more striking when rates in the western division are combined with those in the south-central and north-central divisions—the two next highest divisons. The western division, which encompassed Arizona, California, Colorado, Idaho, Montana, Nevada, New Mexico, Oregon, Utah, Washington, and Wyoming, can be united with the south-central and north-central because these divisions also included a number of states and territories generally thought of as western in culture and outlook during most of the nineteenth century. The south-central included Indian Territory, Oklahoma Territory, and Texas, while the north-central was composed of Indiana, Illinois, Iowa, Kansas, Michigan, Minnesota, Missouri, Nebraska, North Dakota, Ohio, South Dakota, and Wisconsin.

Undoubtedly, some Americans interpreted high rates of divorce in these states and territories as a temporary development resulting from stresses of migration and settlement. But Commissioner Wright pointed out that the upwardly spiraling divorce rate was a western rather than a frontier phenomenon. Although it was reasonable to expect the divorce rate to stabilize as western areas became more settled, Wright noted "no such tendency" was "apparent in the figures for divorce, and in fact an opposite tendency" appeared "to be at

*Divorces Per 100,00 People*

|      | North Atlantic | South Atlantic | North-Central | South-Central | Western |
|------|----------------|----------------|---------------|---------------|---------|
| 1870 | 26 | 8 | 43 | 18 | 56 |
| 1880 | 29 | 13 | 56 | 37 | 83 |
| 1890 | 29 | 21 | 73 | 63 | 106 |
| 1900 | 39 | 33 | 95 | 97 | 131[3] |

work."[4] In his view, high rates of divorce in the West seemed likely to continue.

Wright reported the "most common single ground for divorce" was desertion. During the period of the first Census Bureau study, 1867–86, desertion was higher nationwide than in the western division, but during the period of the second study, 1887–1906, the western desertion rate exceeded the national average.

The rise in desertion cases in the western division may have been caused by a shift in population. Between 1867 and 1886, abandoned spouses in eastern and central states obtained divorces after their spouses went westward. In 1840, for example, Anna Tucker Morrison of Mobile, Alabama, left her husband and relocated in Jacksonville, Illinois. He later obtained a divorce in Alabama on the ground of her desertion. After 1887, however, far more people, and far more potential deserters, lived in western states and territories. Many abandoned spouses were westerners and obtained divorces in western jurisdictions. One such case was deserted Oklahoma Territory husband who sought a divorce on the ground of his wife's prolonged absence, explaining that she had "gone over-land in a wagon west to some Western state."[6]

A sizable number of western settlers obtained divorces because their spouses refused to migrate with them. A wife who remained in a couple's former home was considered a deserter in jurisdictions whose laws stated that a husband's domicile constituted the family domicile. A typical case was that of Berne Ball. In 1895, he divorced his wife of twenty-eight years on the ground of desertion after she refused to migrate from New York City to Logan County, Oklahoma.[6]

Desertions that never reached divorce courts also appear to have been rampant throughout the West. But, because no one was counting, it is impossible to measure the extent or duration of desertion.

Diaries and letters disclose many instances of desertion. One of these took place in Iowa during the early 1830s. Diarist Caroline Phelps described her sister Eliza's disillusionment with her "dissipated" husband. Eliza and her daughter occasionally moved in with Caroline and her husband. Finally, Eliza's husband appeared at Caroline's home where, according to Caroline, he "cryed like a child,"

*% of Divorces Based on Desertion*

|  | *Nationwide* | | | *Western Division* | | |
|---|---|---|---|---|---|---|
|  | Total | Men's | Women's | Total | Men's | Women's |
| 1867–1886 | 38.5 | 45.7 | 34.7 | 30.4 | 47.7 | 23.1 |
| 1887–1906 | 38.9 | 49.4 | 33.6 | 39.6 | 66.3 | 29.5[5] |

gave their daughter "some apples & Candy," and told Eliza "to take good care" of the child. He then boarded a Mississippi River steamboat bound for New Orleans and was never heard from again.[8]

In another case, Mary Jane Megquier left her husband, Dr. Thomas Lewis Megquier of Winthrop, Maine, and relocated in California during the early 1850s. For several years, she ran a boarding-house. When Thomas died in 1855, she lingered on in California, while reconciling with her daughter Angie by mail. In an 1856 letter, Mary Jane said that she was pleased that Angie did not think her "an indifferent wife and Mother" for leaving a man who was draining her health and well-being.[9]

Desertion continued to be widespread during the latter part of the nineteenth century as well. "Wanted" posters show that both husbands and wives deserted. Often, abandoned mates wanted deserters to return, even though the circumstances surrounding abandonment made it clear that the deserters had little interest in doing so.

An especially poignant "wanted" poster was issued by the wife of Walter Davis of Sherman, Texas. It began with an eye-catching plea: "For Humanity's Sake, Help Me Locate Walter L. Davis." She then told a strange tale about his disappearance. From his letters, she had learned that he had been kidnapped and was being forced to "work in some mine unknown to him." In these letters, which were mailed "on the trains between Fort Smith, Arkansas, and Oklahoma City" and "arrived tolerably regular," Davis explained he was unable to send his address because he was unaware of his location. She concluded by offering a reward for information about Davis, a sum to be paid by Davis's father.[10]

An abandoned Oklahoma Territory husband's poster described another kind of abduction. He believed a well-built, dark-haired, mustachioed hypnotist had "exercised his power" over his wife and spirited her away. He offered a liberal reward for information about his attractive, well-dressed, and healthy-complexioned wife and the spell-casting hypnotist.[11]

Most police posters lacked such drama. Frequently, wives were looking for information regarding absent husbands. A Mrs. Robert Archibald of Elgin, Illinois, listed her fifty-two-year-old husband among the missing and offered a $100 reward for his return. The wife of forty-one-year-old Peter Hempler of Chicago, did more than just state her plight. Annie Hempler also swore out a warrant for "wife and child abandonment" and offered a "liberal reward" to the person who apprehended her husband. Of course, some women were too poor to offer rewards. The wife of thirty-year-old Frank Limbrack of Owensboro, Kentucky, was searching for her husband because she and her children were "in destitute circumstances," while the wife of forty-one-year-old George W. Roach of Harrisburg, Illinois, explained her poverty occurred overnight: her husband took $1400, "every cent of money she had," with him when he deserted.[12]

Husbands also sought information concerning missing wives. Twenty-two-year-old Mrs. May Walker of Pueblo, Colorado, was said to have disappeared in the company of seventeen-year-old Mabel Wade, a "very good looking girl." And S. D. Gilbert of Smithville, Texas, offered a reward of $50 for the return of twenty-two-year-old Carrie Gilbert after she left town suspiciously dressed in a "long grey cloak, hat and heavy black veil."[13]

Such cases remind us once again that divorce is only one indicator of marital dissatisfaction. Desertion indicates many disgruntled spouses ended their marriages by disappearing rather than by divorcing.

**Trends and Patterns**

According to the Wright report, cruelty was the ground growing fastest in popularity, especially among western women.

The growing use of cruelty as a ground for divorce may have resulted, at least in part, from the inclination of legislators to extend its scope. Before the Civil War, many northeastern and southern lawmakers expanded the definition of cruelty to include verbal abuse. During the latter part of the nineteenth century, western lawmakers also added verbal abuse to cruelty provisions. In 1877, for example, legislators in Dakota Territory broadened the ground of cruelty to include "mental suffering." In 1890, Oklahoma territorial legislators did so as well.[15]

Still, a random sample of Oklahoma Territory divorce cases from the 1890s reveals that most petitioners cited a combination of physical and verbal abuse. Perhaps both forms of mistreatment existed in their marriages, or perhaps these petitioners were hesitant to take a chance on the new mental suffering provision. In 1890, for example, Mary Ames testified that her husband kicked and struck her. He also reportedly spurned her "overtures," thus making "the relations of man & wife a mockery and something to be dreaded." Two years later, Vera Wilder accused her husband, Horace, of extreme cruelty.

*% of Divorces Based on Cruelty*

| | Nationwide | | | Western Division | | |
|---|---|---|---|---|---|---|
| | Total | Men | Women | Total | Men | Women |
| 1867–1888 | 15.6 | 5.4 | 21.0 | 19.8 | .09 | 23.0 |
| 1887–1906 | 21.8 | 10.5 | 27.5 | 21.8 | 13.1 | 24.2[14] |

He was, she stated, a habitual drunkard who physically abused her. She testified he also called her "a chippy in presence of others (meaning a hoar)" and said that she was "a dam old hoar." In 1895, Alma Winston accused her husband of beating her and directing obscene language at her. And in 1897, Manie Brown stated her husband was a drunkard who struck her and called her "God Damned Whore, God Damned Bitch, and God Damned Thief."[16]

Like women petitioners, men who accused their mates of cruelty cited both physical and verbal mistreatment. In 1895, J. Dayton Thorpe testified before an Oklahoma territorial court that his wife, Abbie, struck and beat him over one hundred times, threw scissors at him, and aimed a revolver at him which she repeatedly snapped to frighten him. He added that she regularly called him a "damned old fool" and a "damn son-of-a-bitch," and told him to "go to hell" on several occasions when he asked her "civil questions." According to him, Abbie finally abandoned him and their daughter, saying that "she did not want the child that it looked too much like its father, she had no use for it." The paperwork ended before a divorce was granted, but it is unclear whether Thorpe dropped his suit or the case was dismissed after Abbie countercharged him with paying "improper attentions" to a hired woman and squandering the $3000 she brought to the marriage.[17]

Commissioner Carroll Wright reported another national trend that existed in intensified form in the West: women obtained more divorces than men. Nationally, women received approximately two-thirds of divorces granted, and men received one-third.[18] In western states and territories, women got at least two-thirds of the decrees and a considerably higher percentage in California, Colorado, Illinois, Indiana, Iowa, Kansas, Michigan, Minnesota, Montana, Nebraska, Nevada, Ohio, Washington, Wisconsin, and Wyoming.[19]

These divorcing women received alimony more frequently than women in northeastern and southern states. Most women who requested alimony were plaintiffs rather than defendants, for it was widely believed that a guilty wife forfeited her claim to a husband's earnings if her misbehavior brought the marriage to an end.

Clear regional differences existed in the application of this principle. Pennsylvania judges ordered alimony in 0.4 percent of divorce cases, Alabama courts in 1.9 percent of cases, and Massachusetts judges in 6.1 percent of cases. But Wisconsin judges ordered alimony in 34.3 percent of cases, and Utah judges did so in 32.1 percent of cases.[20]

Even when a man initiated the divorce and proved his wife guilty, judges sometimes awarded alimony to the female defendants. One such case was that of Lorenzo B. Lyman who filed for divorce from his wife, Fannie, on Aug. 30, 1891, in Oklahoma Territory. Although she was living in Montana, she quickly arrived at the Guthrie courtoom, where she testified that Lorenzo had refused to support her and her two children, one of whom was an epileptic. She had supported herself and the children by taking in boarders and running an employment agency. Fannie countersued, charging abandonment and asking for custody of the children, a $2000 lump sum settlement, and $600 per year. The court found for her and awarded her custody, alimony, and a portion of the couple's joint property. In another case, David Hughes sued his wife, Mary, on the ground of adultery in 1897, also in Oklahoma Territory. Mary appeared in court where she denied the charge, asked for $50 per month alimony, and gave the court a detailed description of two amputations of her arm, a result of David shooting her "without any cause." David got a divorce, but Mary received $1000 alimony (lump sum) and $50 in attorney's fees.[21]

Generally, alimony awards were insufficient to women's needs, especially women who received custody of children. Presumably women who had worked during their marriages would continue to do so after they divorced. In 1891, for example, Susie Gleason told an Oklahoma territorial court she had been "compelled to support herself" because her husband drank excessively and failed to provide; Lina Weyach said she worked as a washerwoman. Later that year, Jennie Zeller stated that she was a seamstress abandoned by her habitual drunkard husband. In 1893, Marian Quick said she had earned her own living and provided "board" for her abusive husband as well. A few days later, Phoebe Wise testified she worked as a hired girl and laundress. In 1895, Grace Rowland and Belle Beck both mentioned supporting themselves after abandonment, but they neglected to specify their jobs. Also in that year, Adolfine Schwart explained that she "washes & sewes for others." In 1899, Eva Bowers noted she had "worked for her own living" because her husband refused to support her. The following month, Martha Condron maintained she supported herself and her alcoholic husband by working as a dressmaker.[22]

Many women who had been unemployed during their marriages took jobs to pay for divorces and to support themselves. One courageous Wyoming woman left her alcoholic gambler husband, taking only $50 and a horse. She traded the cash and animal for a cabin and a small piece of land that she farmed to support herself and her five children. She later worked as a postmistress, traveling peddler, shopkeeper, and rancher. Other women also found whatever employment they could. A young Nebraska woman worked at a variety of jobs in the town of Hastings, while another filed on a homestead claim in western Nebraska. Several other aggrieved wives worked as domestics and seamstresses. One gave piano lessons. Another ran a farm; she hauled hay, cut oats, and "thrashed." And another loaded her eight sons into a covered wagon, traveled from Ohio to Illinois, and established a farm.[23]

Western divorce-seeking women also had a slightly better chance of getting child custody than did women in southern and northeastern states. Between 1887 and 1906, Census Bureau figures show that women received custody at a three-to-one ratio to men. In western states, women received custody at a noticeably higher ratio, especially in California, Colorado, Idaho, Illinois, Indiana, Iowa, Kansas, Michigan, Minnesota, Montana, Nebraska, Nevada, Ohio, Utah, Washington, and Wisconsin.[24]

When women were denied custody, public sentiment was often on their side. In 1906, for example, Evelyn Blakeney kidnapped her seven-year-old daughter, locked herself and her daughter in a lavatory on an eastbound Rock Island train, and fired a bullet through the door when the sheriff tried to break in. After her arrest, a scandalous hearing ensued, in part because her former husband was a well-known attorney who was a candidate for delegate to the Oklahoma Constitutional Convention. Throughout, she was supported by a condolent crowd and sympathetic newspaper coverage.[25]

Obviously, western divorce was far from a nirvana for women. Alimony awards were often small, one-time sums, and child custody decisions were erratic. Consequently, many women remained in difficult marriages because they were unable to support themselves and feared losing their children. One especially touching case was that of Lena Tow, who left her husband and took their three small children to Montana. After her borrowed funds ran out and she failed to find a job, she dejectedly returned to her husband in Norway, Iowa. A despairing Kansas woman took another tack; she simply waited for her verbally abusive, alcoholic husband to die.[26]

Yet another national trend found in the western states and territories was the spread of divorce—meaning decrees granted by white legislatures and courts—to virtually all groups of people. For instance, it appears that African Americans increasingly used white laws rather than their own churches or community customs to obtain divorces. Although court records seldom specified the race of petitioners, newspaper accounts of divorces often did so. In 1893, a Guthrie, Oklahoma, newspaper noted that Aaron Jordan divorced his wife, Sarah, because she refused to leave Arkansas and join him in Oklahoma; both were African-American. A few years later, a Vinita, Oklahoma, paper reported that Laura Brookins divorced her husband because of his abusive behavior; again, both were African-American.[27]

Newspaper accounts also reveal that interracial marriages occasionally ended in divorce. In 1908, a white woman, Alexandria Lewis, sued her African-American husband, Albert, for divorce on the ground of misrepresentation. She explained that Albert had represented himself as an Osage Indian. After she discovered he was an African American, she requested a divorce. Although they had only been married for four months, she asked for $6 per week alimony.[28]

American Indians also adopted the white practice of obtaining divorces through legislatures and courts. A case in point was the Cherokee Indians, who adopted white-style divorce along with numerous other aspects of white civilization. Articles in the *Cherokee Advocate* and records of the Cherokee National Council during the 1880s indicate that the Council enacted several "Acts Granting Divorces"—in other words, legislative divorces. After the passage of an 1881 divorce statute, Cherokee circuit and district courts granted judicial divorces in Indian Territory, later part of Oklahoma Territory. Grounds for divorce were adultery, imprisonment of three years or more, wilful desertion for one year, extreme cruelty by violence or other means, and habitual drunkenness.[29]

After Oklahoma Territory was formed in 1890, Indian couples increasingly availed themselves of white divorce courts. When Luke Bearshield divorced his wife, Nellie, in 1893, a Guthrie newspaper declared he was the first Indian to obtain a "legal"—meaning, white—divorce. During subsequent years, other Indians sought divorces from white courts. One of these involved the common law wife of Eastman Richard, who petitioned for a divorce, custody of their two children, and $40,000 alimony.[30]

People representing all jobs, professions, and social classes also requested divorces from civil courts. From incomplete figures, Commissioner Wright concluded that working-class people divorced more than middle- and upper-class people. Also, actors and showmen obtained the highest proportion of divorces, while agricultural laborers and the clergy obtained the lowest. Divorce case records from Oklahoma Territory during the 1890s show that the jobs held by male divorce-seekers ranged from homesteaders to well-to-do merchants, while those held by female petitioners included housewives, domestic workers, boardinghouse-keepers, and milliners. Newspaper accounts from the 1890s indicate that divorce also touched the lives of professional and business people: a professor, two judges, a colonel, a "wealthy German," and a district leader of the Republican party.[31]

The ethnicity of divorce-seekers remains unknown because neither divorce petitions nor court documents recorded ethnic origins of litigants. Newspapers also took little note of ethnicity; the account that mentioned a wealthy German was unusual.

## The Divorce Mill Panic

By the late nineteenth century, many Americans were disquieted by divorce in general and western divorce in particular. In their eyes, it seemed the West was rapidly leading the nation into moral and social decline. Their fears were reinforced by frequent reports of outrageous western divorce trials and of free-wheeling divorce mills that granted quick divorces to migratory divorce-seekers from stricter jurisdictions.[32]

Divorce mills alarmed many Americans. Some people recognized the injustices that could occur in quick and easy divorces granted to out-of-state divorce-seekers. Others were embarrassed because their home states and territories were widely censured for their permissive divorce laws and procedures. And others reacted to divorce mills with consternation because they saw them as harbingers of greater evils: the spread of immorality, a breakdown in family life, and the fall of American society.

During the latter part of the nineteenth century, divorce mills fueled the divorce debate and fostered a widespread demand for change. In 1885, Samuel W. Dike, secretary of the National Divorce Reform League, pointed out that divergent divorce laws made divorce mills and migratory divorce possible. "The divorce broker," he wrote, "sits in his office, and from the compilations prepared for his use, assigns his applications to one State or another as may best suit each case."[33] Consequences of divorce mills, whether real or assumed, primed many people to respond with indignation to the divorce mill scenario.

Certainly, during the closing decades of the nineteenth century, the rapid growth of divorce stunned many people: one of fourteen to sixteen marriages ended in divorce in the United States during the 1880s. In addition, divergent divorce laws caused confusion and sometimes corruption as well. In 1888, a New York attorney compiled a chart showing that all states and territories, except South Carolina, allowed divorce. Of the forty-seven states and territories surveyed, forty allowed the inclusive, flexible ground of cruelty. Numerous other grounds were available, and even anti-divorce South Carolina provided several grounds for annulment.[34]

The following year, Margaret Lee's best-selling novel, *Divorce,* was republished with an ominous prefatory warning about divorce in the United States. Written by Britain's Prime Minister, W. E. Gladstone, the book's preface described a "social revolution" in the family and asserted that in the United States the "marriage controversy" had "reached a stage of development more advanced than elsewhere." According to Gladstone, one of ten Connecticut marriages and one of seven California marriages ended in divorce. He concluded, "America is the arena in which many of the problems connected with the marriage state are in course of being rapidly, painfully and perilously tried out"; the nation would "do much by her example to make or mark" the destinies of other countries.[35]

Western divorce mills seemed to be the height of laxity and permissiveness: the ultimate inducement to divorce-seekers to flee strict laws in their home states and seek a divorce in more lenient jurisdictions. Consequently, divorce mills elicited impassioned criticism and indignant responses.

During mid-century, Utah was branded a divorce mill as a result of Mormon policies concerning marriage and divorce. The Church of Jesus Christ of Latter-day Saints was founded by Joseph Smith in Fayette, New York, in 1830, but its members, commonly known as Mormons, fled persecution by moving to Ohio, Missouri, and Illinois. It was at Nauvoo, Illinois, on July 12, 1843, that Smith received a revelation saying that Mormons must practice polygamy—meaning that one husband wed several wives. This innovation drew enormous enmity from outsiders; in 1844, an anti-Mormon mob lynched Joseph Smith.

After this calamity, thousands of Mormons trekked to a desert in Utah that lay outside the boundaries of the United States. They hoped to live in peace, free from persecution and regulation by laws stipulating that marriages be monogamous. Under the leadership of Brigham Young, Mormons established Salt Lake City in 1847 and the state of Deseret in 1849. In 1850, the United States Congress recognized Deseret as the Territory of Utah, which brought Mormons back within the jurisdiction of the United States. Brigham Young served as governor of the new territory until 1857, when conflicts with the United States government, largely over polygamy, ended his tenure. Although the U. S. Congress enacted anti-polygamy statutes in 1862, 1882, and 1884, church officials refused until 1890 to abandon the practice.[36]

During these years, many Americans harshly criticized Mormon practices, for they saw polygamy as a threat to long-held and widely cherished conceptions of marriage. In 1850, John Gunnison, an army officer stationed in Salt Lake City, wrote his wife that "some things happen in this polygamy loving community which would astonish the people in the States." He added that it was easy to see "the influence of polygamy in degrading the female sex." Some years later, another anti-polygamist, Philip Van Zile, thought about running for Congress so he could "do this country good" by eradicating "that relic of barbarism from its fair name."[37]

In addition to polygamy, the divorce practices of the Latter-day

Saints shocked Gentiles, as Mormons called non-Mormons. Beginning in 1847, Mormon church leaders regularly granted divorces. Because they lacked the legal power to terminate marriages, they claimed they limited themselves to divorcing polygamous couples whose marriages fell within the jurisdiction of the church. Brigham Young reportedly granted over 1600 divorces during his presidency of the church between 1847 and 1876. Although Young theoretically opposed divorce because it contradicted the Mormon belief in eternal marriage, he was willing to terminate contentious and other unworkable marriages. On one day, he relieved George D. Grant of three wives and a few weeks later, parted him from a fourth.[38]

Young personally lacked sympathy for men such as Grant: "it is not right for men to divorce their wives the way they do," he stated in 1858. He had slightly more compassion for women. Although he often counseled a distraught wife to stay with her husband as long "as she could bear with him," he instructed her to seek a divorce if life became "too burdensome." In 1861, Young instructed husbands to release discontent wives.[39]

As news of Mormon church divorces reached the Gentile world, public outrage against Mormons flared. After 1852, when the first Utah territorial legislature adopted a statute that permitted probate courts to grant divorces, many people became highly critical of lenient civil divorces as well.

The 1852 Utah Territory statute was objectionable because in addition to listing the usual grounds of impotence, adultery, wilful desertion for one year, habitual drunkenness, conviction for a felony, inhuman treatment, it included an omnibus clause. According to this clause, judges could grant divorces "when it shall be made to appear to the satisfaction and conviction of the court that the parties cannot live in peace and union together and that their welfare requires a separation." In addition, the 1852 statute contained a loose residency requirement: a court need only be satisfied that a petitioner was "a resident of the Territory, or wishes to become one."[40]

As a result of the 1852 statute, civil divorces were easy to obtain in Utah Territory; a couple could even receive a divorce on the same day they applied for it. Unlike most other jurisdictions, Utah judges accepted collusion—an agreement to divorce between husband and wife. A married couple could appear in court, testify that they agreed to divorce, and receive a decree. Records of the Washington County probate court between 1856 and 1867 contain several such cases. On February 12, 1856, John and Sarah Wardall petitioned for divorce and requested equal division of their children and property. The judge agreed: John received custody of the two oldest boys and Sarah got custody of their daughter and youngest boy. The Wardalls also amicably split two beds, four pillows, two bolsters, two blankets, and other household equipment down-the-middle. What could have been a difficult divorce turned out to be an administrative matter completed in a few minutes.[41]

In an unusual case of mutual agreement, a woman's father appeared before a Washington County judge. He testified that his daughter and her husband had asked him to apply for a divorce on their behalves. The judge, who knew the couple, stated that husband and wife wanted to divorce so that they could "marry whomsoever they will or can." Because he believed that mutual agreement resulted "in the most good to both Parties," he granted the divorce. It became final four days later when the couple submitted a property settlement.[42]

When Jacob Smith Boreman, a non-Mormon from Virginia, became United States district court judge in the Salt Lake City region in 1872, he was shocked by Utah divorce laws and procedures. Boreman was especially surprised that judges accepted collusion and that divorce-seekers could file petitions, enter proof of grounds, and receive divorce decrees "all on the same day." Boreman remarked that such practices "made it no difficult matter to secure a divorce in a probate court," especially when most judges "had no legal training, but on the contrary were densely ignorant of the rules of law."[43]

Boreman himself heard a portion of one of the most dramatic divorce cases in Mormon history. In 1873, Ann Eliza Webb Young brought suit against her husband, Brigham Young. Young seemed willing to divorce Ann Eliza, but unwilling to pay the requested alimony: $20,000 costs plus $200,000 to support Ann Eliza and her children. Young, who had once offered to divorce any wife who wished to leave him, fought Eliza's petition by arguing that their marriage was illegal because it was polygamous, thus unrecognized by United States law. According to Boreman, Young believed that if he won, he would be free from alimony; if he lost, polygamous marriages would have garnered legal recognition, for if a judge gave Ann Eliza a divorce he would have also inadvertently declared the Young's polygamous marriage valid.

After an 1874 federal law moved Utah divorce cases from probate to district courts, Boreman became the presiding judge in the case of *Young v. Young*. Boreman ordered Brigham Young to pay temporary alimony to Ann Eliza, but he had to imprison Young to make him pay. The suit was dismissed in 1877 by another district court judge who refused to recognize Brigham Young's polygamous marriage to Ann Eliza. Consequently, Ann Eliza Young failed to get a divorce decree and alimony, while Brigham Young failed to get recognition of polygamous marriages.[44]

Despite its lenient divorce laws, it is unclear whether Utah was a divorce mill. Between 1867 and 1886, Utah courts granted 4,078 divorces. Of these, 1,267 couples had married in Utah. It is impossible to know how many of the remaining 2,811 cases involved migratory divorce-seekers or those who were converts anxious to join the Latter-day Saints after they freed themselves from unwilling mates. High migration rates into Utah during these years, however, suggest that most divorces were probably obtained by would-be converts rather than migratory divorce-seekers.[45]

During the early 1890s, divorce mills allegedly emerged in the Dakotas. Members of the House of the Dakota territorial legislature introduced divorce into the region in 1862 when they granted divorces to Minnie Omeg from her husband, C. Omeg, and to General William Tripp from his wife, Sarah A. Tripp. Other divorce-seekers soon requested legislative divorces.

Legislators felt harried and distracted by these petitions and complained that they diverted time and energy from other business. Thus, in 1865–66, the legislature shifted primary jurisdiction for divorce to the courts. In the first territorial divorce law, they also established adultery and life imprisonment as grounds for divorce and required that petitioners be residents of the territory when the adultery or imprisonment occurred. In 1866–67, the legislature specified seven grounds for divorce, set the residency requirement at ninety days in the country where the suit was filed, and recognized the practice of notice by publication. In 1877, legislators reduced grounds to six, but added the flexible ground of neglect and broadened the plea of cruelty to include "mental suffering." In addition, a new clause stated that the ninety-day residence could be established in any county. These provisions remained in force throughout the 1880s.[46]

During the 1880s, the territory experienced an economic boom and increased immigration. Consequently, in 1889, an enabling act divided Dakota Territory into two new states, North Dakota and South Dakota. South Dakota's legislature retained the six grounds for divorce and the ninety-day residency requirement.[47]

After 1889, South Dakota courts reportedly granted a growing number of divorces. The bustling city of Sioux Falls quickly garnered a reputation as the newest divorce mecca in the United States. Located on the eastern border of the state, Sioux Falls was the hub of major railroad lines which transported out-of-state divorce-seekers. As a commercial center, Sioux Falls had an abundance of attorneys and courtrooms. It also offered increasingly sophisticated theater, shopping, lodging, and recreational facilities. By 1893, the *Evening Argus Leader* remarked that Sioux Falls was getting "metropolitan with a vengeance;" it encompassed ten gambling halls, thirty-seven "holes in the wall," and one hundred prostitutes.[48]

Sioux Falls was widely viewed as a rollicking and extremely lenient divorce colony. Charles H. Craig, editor of the Sioux Falls *Daily Press*, supposedly dubbed the city a divorce colony; other journalists supplied lurid details about it. In 1895, a Chicago journalist and an attorney visited Sioux Falls briefly, then wrote a titillating book about the city. According to them, the "Queen City" prospered "through the misery, sin, and follies of wives and husbands." A few years later, journalist George Fitch expanded this portrait in the *American Magazine*. According to Fitch, Sioux Falls boasted many "corpulent homes built from the proceeds of the divorce industry." At any time, one to five hundred divorce-seekers from "East and West, Canada and foreign lands" were seeking divorces in Sioux Falls. Fitch estimated that the Minnehaha County court in Sioux Falls granted seventy-five to one hundred divorces each year.[49]

Other accounts further inflamed people's imaginations. In 1908, a novel by a Sioux Falls dentist characterized the city as the "scene of an endless social drama, the battle-ground upon which a giant domestic problem is daily fought." The following year, Jane Burr's "letters" described the experiences of a female divorce-seeker living alone in Sioux Falls.[50]

As the city's notoriety grew, many citizens of Sioux Falls became vehement opponents of the local divorce industry. While some attorneys, hotel-keepers, and businesspeople thought that lenient divorce laws helped the local economy, other inhabitants railed against these laws. William Hobart Hare, Episcopal Bishop of South Dakota between 1873 and 1909, emerged as leader of the opposition. Viewing women and men as incomplete beings comprising a "whole" only in marriage, Hare rejected divorce as a solution to family disharmony. He especially attacked merchants and attorneys who, in his view, "capitalized on unhappiness" by soliciting "divorce business."[51]

Bishop Hare began his anti-divorce campaign in early 1893 with a stinging pastoral. He argued that ease of divorce and remarriage was causing the nation to drift into "consecutive polygamy." Hare also indicted South Dakota for providing "an impetus to this downward progress by the laws on divorce which we allow to remain upon our statute book." Hare added that he was ashamed to recruit female students for the All Saints School that he had established in Sioux Falls because the town's environment was tainted by divorce.[52]

Hare's stand on the divorce issue gained support from several quarters. In 1893, for example, an editorial in the *Nation* applauded him for opposing "scandalous divorce laws" and called for "every encouragement" of his work. The editorial also disparaged the "jocular treatment of the [divorce] scandal in the newspapers."[53]

The South Dakota legislature began considering divorce reform in

1890. Legislators discussed a reform bill, but it never reached the floor of the House. In 1893, the legislature passed a bill requiring six months' residency in the state before divorce proceedings could be initiated and a one-year residency before a final decree could be awarded. When this moderate change failed to satisfy critics, legislators introduced several stricter residency bills during subsequent sessions, but none became law.[54]

In 1907, a one-year residency requirement became the focus of debate, political maneuvering, and lobbying. Bishop Hare, who had recently served as a delegate to a national conference on divorce law, sent the conference's recommendations, including a one-year residency requirement, to the South Dakota legislature. When the one-year bill came up for vote, both houses passed it by unexpectedly large margins. The following year a popular referendum also gave it strong support. The measure went into effect on January 1, 1909. The linchpin of South Dakota divorce mills—short residency—was gone.[55]

It is unclear whether some South Dakota cities were divorce mills that granted many quick divorces to out-of-state petitioners. One commentator pointed out that the divorce rate in Minnehaha County in 1903 was extremely high: one divorce per 2.26 marriages compared with a national average of one divorce per seven marriages. Thus, he argued, it must have included many migratory decrees. But another analyst maintained that "the volume of business was certainly greatly exaggerated by the divorce colony propagandists." The latter interpretation is supported by Census Bureau statistics that reveal a much lower divorce rate than reported by South Dakota newspapers at the time. These figures also indicate that the divorce rate in such other central states as Kansas, Arkansas, and post-divorce-reform Indiana surpassed South Dakota's allegedly high rate.[56]

Despite the lack of satistical proof, the image of Sioux Falls as a divorce mecca continues to loom large in people's minds. Looking back on her childhood in the city, one woman remembered "hundreds" of rich, glamorous, and self-indulgent divorce-seekers. In 1949, a history of Minnehaha County included a section on "divorce days" and the "divorce colony." And as recently as 1983, the Sioux Falls *Tribune* ran stories depicting the city as an 1890s divorce mill.[57]

Between 1893 and 1899, South Dakota's northern neighbor experienced a similar, but more short-lived, phenomenon. North Dakota had shared territorial legislation with South Dakota, including a ninety-day residency requirement for divorces. This provision was later defended by Charles A. Pollack, a former Iowan who became district attorney of Cass County, North Dakota, in 1884, and served as a judge on the Cass County bench between 1897 and 1917. Pollack denied that the short residency requirement was intended to draw divorce business to North Dakota. He explained that short residency requirements were common in frontier regions; they were intended to lure settlers by allowing doctors and lawyers to practice on arrival and men to vote after thirty days' residence. Pollack argued that a ninety-day residency requirement for divorce was restrictive by contrast. He added that at the time of its adoption, no one foresaw that divorce-seekers would take advantage of the short residency "in such large numbers that it became a burning disgrace to the state."[58]

When North Dakota became a state in 1889, it retained territorial statutes stating that divorce would be a judicial matter and that a divorce petition could be initiated after ninety days' residency in the state. An 1899 statute listed seven grounds for divorce, including the catch-all grounds of cruelty and neglect. Because North Dakota was less accessible than South Dakota, these provisions failed to attract divorce-seekers. But, when the South Dakota legislature lengthened the state's residency requirement from three to six months in 1893, some of the divorce business reportedly shifted northward.[59]

Fargo, the seat of Cass County, especially attracted divorce-seekers. This growing city on the eastern border of North Dakota offered all the necessary facilities: major railroads, hotels, shops, entertainment, attorneys, courts, and newspapers.[60] By the mid-1890s, Fargo's reputation as a divorce mill rivaled that of Sioux Falls.

This development astounded and appalled many of Fargo's citizens. The Women's Christian Temperance Union especially opposed North Dakota divorce statutes. Elizabeth Preston, president of the North Dakota WCTU between 1893 and 1901, objected to the speed of easy divorce. Preston termed the ninety-day residency clause "odious" and worked energetically for its repeal. John Shanley, first Roman Catholic Bishop of Fargo, was another energetic critic, opposing divorce in general and the form it took in North Dakota in particular. During the mid-1890s, Shanley crusaded for tighter controls. He later explained that his "fierce and successful warfare" against permissive divorce laws gained him "undeniable unpopularity" in Fargo, especially among tradespeople, attorneys, and journalists, who welcomed the economic "bonanza" easy divorce created.[61]

A growing number of North Dakota legislators agreed with these critics. In 1893, 1895, and 1897, the legislature considered, but failed to pass, one-year residency bills. In 1899, the legislature approved a one-year residency requirement that became law on July 1, 1899.[62]

In mid-1899, then, the divorce days of North Dakota came to a halt. But since that time, writers have repeated and enlarged the tale of booming divorce mills in Fargo and other North Dakota cities. One sociologist computed an extremely high divorce rate from unidentified sources. A historian also neglected to cite evidence for his assertion that divorces ran "into several thousand," and "one Fargo judge heard 350 divorce cases in a single year."[63]

Census Bureau statistics fail to confirm North Dakota's divorce mill image. Although one historian of divorce asserted that "Fargo's divorce business was probably greater than Sioux Falls' had been," government figures indicate Fargo courts granted fewer divorces than courts in Sioux Falls. Cass County court "register of actions" books also indicate numbers of divorces granted in Fargo were less than those in Sioux Falls.[64]

Evidently, Fargo's divorce mill image resulted more from contemporary newspaper accounts than statistical reality. In 1896, for example, the St. Paul, Minnesota, *Dispatch* reported on hearsay that 150 migratory divorce-seekers resided in Fargo. Other newspapers frequently featured migratory cases. One of these was the 1899 divorce hearing of George Van Vleck, a petroleum entrepreneur. Van Vleck established residence in Fargo and hired a North Dakota attorney, while his wife remained in Buffalo and hired a New York lawyer. When the Cass County court granted the divorce, the Fargo *Forum* declared, "Van Vleck has been a resident of Fargo for many months and made many friends here who will congratulate him on securing the decree."[65]

During the 1890s, Oklahoma Territory was supposedly yet another western divorce mill. In 1890, the framers of the Organic Act acted in haste and simply copied Nebraska's divorce provisions. Realizing that Nebraska statutes were unsuited to Oklahoma Territory, they within the year created territorial divorce statutes. In one section, the new territorial statutes lodged jurisdiction for divorce in district courts and required that petitioners live in the territory for ninety days before applying for divorce. In another section, the statutes gave jurisdiction to both probate and district courts and required that a petitioner live in the territory for two years and in the county of filing for six months before applying for a divorce.[66]

The 1890 statutes also allowed notification of proceedings by publication and provided for appeal of divorce decrees. Grounds were adultery, extreme cruelty, wilful desertion, wilful neglect, habitual intemperance, and conviction for a felony. The statutes stipulated that extreme cruelty meant "the infliction of grievous bodily injury or grievous [*sic*] mental suffering upon the other, by one party to the marriage."[67]

In 1893, the Oklahoma territorial legislature revised and clarified these contradictory statutes. The new statute lodged jurisdiction for divorce exclusively in the district courts, set residency at ninety days, and specified ten grounds, including the pliant category "gross neglect of duty." To stop immediate remarriages, a new provision stated that a divorce decree would take effect six months after it was granted.[68]

Despite these changes, confusion continued to prevail in territorial court proceedings. Since 1890, probate court judges had been hearing divorce cases after ninety days' residency by the petitioner rather than two years' residency as required by the 1890 law. Although the 1893 reform should have closed down the probate divorce business, probate court judges continued to grant divorces and require only ninety days' residency.

Territorial newspapers maintained that Oklahoma was turning into a divorce mecca. In 1892, the *Daily Leader* reported that divorce-seekers streamed into Oklahoma by the hundreds to take advantage of quick and easy probate divorces. Two years later, the *Daily Times Journal* asserted that the Territory's divorce colony was growing in spite of the recent law prohibiting probate courts to grant divorces.[69]

The legality of probate divorces was tested in an 1894 divorce appeal heard by Judge Frank Dale, chief justice of the Oklahoma Territorial Supreme Court. In *Irwin v. Irwin,* Dale ruled that probate courts lost the power to grant divorces after August 14, 1893. Because Dale's decision invalidated probate divorces granted after that date, Associate Justice Henry W. Scott voted against it. Scott warned that Dale's decision would "make innocent people guilty of adultery and bigamy."[70]

Public alarm flared when the Kingfisher *Free Press* predicted that hundreds of divorces would be nullified, while the Guthrie *Daily Leader* declared that those who had remarried after receiving a probate divorce after August 14, 1893, were now bigamists. To alleviate these problems, the territorial legislature passed an act legalizing divorces granted by probate courts before February 28, 1895.[71]

Despite the confusion, Oklahoma newspapers reported that divorce-seekers continued to pour into the territory. In June 1895, the Edmond *Sun Democrat* speculated that the Oklahoma divorce colony numbered at least five hundred people. The following year, the Alva *Republican* estimated that 2000 divorce-seekers were in the territory.[72]

During these years, Guthrie emerged as the center of Oklahoma's divorce trade. Like other divorce meccas, Guthrie appealed to divorce-seekers because of its convenient railroad lines; lodgings and other amenities; and abundance of lawyers, courts, and newspapers. Anonymity was also possible among its highly mobile and rapidly increasing population. Profit-minded hotel-keepers, attorneys, and others were quick to advertise Guthrie's inducements in eastern newspapers. One attorney even described Guthrie as the ideal divorce resort: scenic, warm climate, financially reasonable, and free of "crowds of loungers and gossips to listen to whatever testimony may be given."[73]

Oklahoma Territory's reputation as a divorce mill provoked criticism inside and outside the territory. In 1893, Governor Abraham J. Seay recommended that the ninety-day requirement be lengthened. Seay pointed out that easy divorce was bringing national shame to the territory. The following year, the Reverend J. F. Palmer of Kingfisher argued that marriage was a religious sacrament rather than a civil construct to be broken at will, while the Philadelphia *Record* lamented that easy divorce was a disgraceful blot on Oklahoma Territory. In 1896, a *New York Herald* reporter accused territorial courts of granting "mail-order" divorces; he published correspondence from an Oklahoma attorney guaranteeing that a petitioner could obtain a divorce without appearing in Oklahoma.[74]

Chief Justice Frank Dale especially denounced attorneys and businesspeople who actively solicited divorce trade. In 1894, he threatened to disbar attorneys who advertised for divorce clients in eastern newspapers and he appointed an investigatory committee to identify such lawyers. Rather than impairing the solicitation of divorce trade, however, Dale's campaign projected Oklahoma further into the limelight.[75]

Other judges opposed easy divorce by refusing to grant divorces that appeared to be migratory. In 1898, for example, the El Reno *News* reported that a Judge Tarsney refused to grant a divorce to a Michigan man on the ground that the man had settled in Oklahoma only to obtain a divorce.[76]

In 1896, Congress intervened; it established a one-year residency requirement in territorial divorce cases. While this bill awaited President Theodore Roosevelt's signature, the El Reno *News* reported that divorce-seekers made a frenzied last-minute dash to submit their petitions before the new law went into effect, a dash reminiscent of the earlier rush of homestead filers at land offices. On May 25, the one-year residency bill became law. Within a few weeks the same newspaper optimistically stated "the Divorce business in Oklahoma has entirely disappeared." But as late as 1908, the *Daily Oklahoman* disclosed that courts were still disputing the precedence of the one-year law over Oklahoma's ninety-day provision.[77]

Was Oklahoma Territory a divorce mill? Despite assertions that the territory "outstripped its competitors in drawing clients," government statistics fail to support this contention. Instead, figures indicate a slightly lower rate of divorce per 1000 population in Oklahoma Territory than in such states as Washington, Montana, Texas, and Indiana.[78]

Although figures indicate fewer divorces were granted in Oklahoma Territory than the press, reformers, and alarmists thought, they fail to show the number of divorces that were migratory. Fortunately, Oklahoma territorial divorce records have been preserved and contain some useful information. A random sample of sixty-six divorce cases, which occurred in Guthrie, Logan County, between 1890 and 1899, disclose that only nine petitioners claimed the minimum ninety-day residency.[79]

Because case records are often sketchy, it is difficult to determine how many of these nine cases were migratory divorces. The first ninety-day case, which was also the first divorce case heard in Logan County, is a good example. *Keller v. Keller* case records indicate the following: the petition was filed on June 17, 1890; the couple had married in 1854 in Pennsylvania; in 1879, Lucinda Keller deserted her husband. In 1890, Frank Keller filed for an Oklahoma divorce that was granted on September 4 after Lucinda failed to appear. But the records do not reveal whether Frank Keller had permanently relocated in Oklahoma Territory or was temporarily residing in the territory to obtain a divorce.

Of the nine cases, five seem to have been migratory. Records of *Zeller v. Zeller,* filed on October 18, indicate Jennie married Harry in 1881 in Illinois, where he abandoned her nine years later. She then moved to Colorado. Because her witnesses were all Colorado residents and she showed no sign of having moved to Oklahoma Territory, the divorce she obtained on December 14 appears to have been migratory.

*Davidson v. Davidson* was filed on May 8, 1895. Anna married Robert in 1884, state not identified, but soon discovered that he was abusive and an alcoholic. After leaving him several times, she brought her children to Guthrie, where she filed a divorce petition. Robert followed her to Guthrie and apparently persuaded her to abandon the suit, for the paperwork stopped shortly after his arrival. Had it been granted, this divorce would have been migratory.

*Bair v. Bair,* filed on May 27, 1895, concerned a couple married in 1894 in Kansas. When Clark Bair discovered his wife, Mary, was pregnant by another man, he left and sought a divorce in Oklahoma Territory. Mary failed to appear and the divorce was granted on October 22. This was a clear-cut case of migratory divorce.

*Mitchell v. Mitchell* was filed on July 26, 1895. Henry and Letitia married in England in 1879 and relocated in Chicago, where she initiated, but neglected to complete, a divorce action in 1892. In contesting Henry's Oklahoma divorce petition, she charged that Henry's residence in Oklahoma was "not in good faith," but only established to pursue a divorce. After the court awarded her temporary alimony, the paperwork came to an end and the divorce action remained incomplete. Based on Letitia's testimony, had the divorce been granted, it would have been migratory.

*Winston v. Winston* was filed on August (no day given), 1895. The

Winstons married in North Carolina in an unspecified year. Edward was suing Alma in Oklahoma for divorce on the ground of abandonment. Alma opposed the suit; she charged Edward with cruelty and having established residency in Oklahoma Territory solely for the purpose of obtaining a divorce. She added that she had sought a legal separation in North Carolina, but that Edward had paid her $50 to withdraw it. Despite her protests, the court granted his divorce request on October 1. Based on her testimony, this was a migratory divorce.[80]

This random sample of Logan County territorial divorce cases does not confirm the existence of a booming divorce mill in Guthrie. The nine ninety-day cases account for less than 14 percent of the total divorce cases sampled. Only a portion of the nine were clearly migratory divorce actions, and not all of them resulted in divorce decrees. In addition, as the 1890s progressed, the number of ninety-day residency claims in the sample decreased sharply. This suggests that the 1896 one-year residency requirement for territorial divorces was generally respected, thus reducing migratory divorce business in the territory.

Nor do such other sources as Guthrie hotel registers support the town's divorce mill image. The guest register of the Hotel Springer for 1892 and 1893 reveals only that a large proportion of guests came from New York, Chicago, and Philadelphia, admittedly jurisdictions with stricter divorce laws than those in Oklahoma. Several of these guests paid for their rooms a month at a time. These people may have been migratory divorce-seekers, or they may have been salespeople and politicians who had long-term business in the territorial capital.[81]

The number of divorces granted to migratory divorce-seekers by western divorce mills was almost certainly exaggerated by contemporary observers.[82] Still, thousands of Americans reacted with alarm to the divorce mills that seemed to litter western areas of the United States during the last half of the nineteenth century. How could state governments restrict and control divorce when western divorce mills tempted divorce-seekers to escape restrictive laws and obtain divorces in more permissive jurisdictions?

The divorce issue was already volatile during the mid-nineteenth century, but the divorce mill panic greatly exacerbated that volatility. In addition, it convinced some people that divorce should be re-examined and cut back, and that "the practice of permitting residence in another state for the purpose of securing a divorce should be abolished."[83]

During the late nineteenth century, many Americans felt the spiraling divorce rate and the rise of divorce mills in western states and territories indicated it was time to re-evaluate divorce in the United

States. They believed these pheonomenon were evidence that such long-held American values as individualism, freedom from tyranny, and a search for personal happiness had flourished in the American West—and had gotten out of hand. Clearly, the time was right for a national call for revision in divorce laws and procedures; the stage was set for the national uniform divorce law movement.

# "Truly the Land Needs a Reform"

☐ In 1879, an editorial in the *New York Tribune* declared that "truly the land needs a reform in the law of divorce." The *Tribune* spoke for the many people who were alarmed by the pervasiveness of divorce in the United States. They believed that divorce was rapidly spreading across the land, affecting poor couples and wealthy ones, disrupting the lives of urban dwellers and extending into the countryside to disrupt the lives of rural couples as well.[1]

What was divorce in the United States coming to? they asked. Had the quest for individualism, freedom, and happiness gone too far? Would rampant instability, gross immorality, a decline of the family, and the destruction of American society result from the growth of divorce?

Many Americans, including members of the National Divorce Reform League, would have answered yes to these questions. But what was the solution? Increasingly, a new breed of divorce reformers, located largely in the northeastern states, blamed the rising divorce rate on divergent divorce laws and argued for uniform national divorce laws. Number of divorces would be reduced if every state and territory enacted identical, stringent divorce laws. And migratory divorce would disappear because strict and lenient jurisdictions would no longer exist. As a result of uniform laws, supporters argued, the alarming divorce rate would stabilize or drop and western divorce mills would disappear.

The seeds of the uniform national divorce law reform movement were planted in 1881, when the New England Divorce Reform League organized under the leadership of Theodore Woolsey, attorney, Doctor of Divinity, and retired president of Yale University. Woolsey's treatise, "History and Doctrine of Divorce," was well known among New Englanders, and his dire warnings against divorce had

108

influenced Connecticut legislators to repeal the state's omnibus clause in 1878. Despite Woolsey's fame, a Congregational minister from Vermont, Samuel W. Dike, soon became the moving force behind the Divorce Reform League. Dike served as secretary of the association from 1884 until his death in 1913.[2]

During the early years, leaders of the New England Divorce Reform League attempted to attract members and to gain insight into divorce reform schemes. When they realized that the League's regional focus hampered achievement of its goals, they pushed for a broader structure. Consequently, the New England Divorce Reform League reorganized on February 16, 1885, as the National Divorce Reform League.

The following year, Secretary Dike's annual report noted that the League's work was both "educational and legislative." Educational endeavors included publications and lectures "to promote public sentiment and legislation on the institution of the Family." Dike, who was very concerned about the welfare of the family, wanted to limit divorce, but not prohibit it entirely.[3]

Dike's 1886 report explained that the League's second area of activity—legislation—promoted stringent divorce statutes like those recently adopted by the legislatures of Vermont, Massachusetts, and Maine. In 1883, for example, Maine had dropped its omnibus clause, tightened residency requirements, and restricted remarriage. Dike was uncertain, however, whether divorce could be reduced by national uniform marriage and divorce laws, an idea that was being widely debated in magazines and newspapers. His report noted, without comment for or against, that writers regularly touted uniform national divorce laws. Also, advocates of such laws supported "the need of uniform divorce laws or of uniformity in the laws of marriage, divorce and polygamy," brought about either by a constitutional amendment or voluntary cooperation among the states and territories.[4]

Dike also made it clear that he believed that a statistical survey of marriage and divorce was a necessary prelude to further modification. Until the proportions of the problem were known, how could anyone launch sensible, effective programs? He commended such states as Maine, New Hampshire, and Massachusetts for initiating state studies of marriage and divorce. He explained that he had already talked with the President of the United States and other government officials "concerning the collection of statistical information in this country and Europe relating to divorce and other matters pertaining to the Family."[5]

Dike actively encouraged individuals and local organizations to send letters and petitions to members of Congress asking them to initiate a study of marriage and divorce. He also urged ministers and ecclesiastical groups to exert their influence. On March 3, 1887, members of both houses of Congress approved a bill appropriating $10,000 to fund a study directed by the Commissioner of Labor to collect "statistics of and relating to marriage and divorce in the several States and Territories and in the District of Columbia." They appropriated an additional $7500 the following year. Commissioner of Labor Carroll D. Wright stretched this minimal appropriation by assigning his regular staff to the project on a part-time basis. As an attorney and statistician who had participated in an earlier, similar study in Massachusetts, Wright had a deep interest in the project.[6]

The first Wright report, as the study results were called, appeared in February 1889. In 1,074 pages, Wright and his team reported on the actions of 2,624 county courts between 1867 and 1886. They omitted only those jurisdictions whose records had been destroyed by fire or loss. This comprehensive report appealed to a wide audience and was reprinted several times.[7]

Advocates of uniform divorce laws were especially anxious to know how divorce statutes affected the divorce rate. The report showed that the divorce rate had declined in three states that had tightened their divorce statutes—Connecticut, Maine, and Vermont—and risen in four jurisdictions that had expanded their statutes—Colorado, Dakota Territory, Massachusetts, and Mississippi. But it also revealed that the divorce rate had climbed steadily in other states that had neither reduced nor enlarged their divorce provisions. Thus, the report concluded, statutes were only one factor among many that influenced the divorce rate.[8]

Uniform divorce law advocates were also intensely interested in knowing what proportion of divorces were migratory. They had long assumed that substantial numbers of divorce-seekers in strict jurisdictions, especially New York and South Carolina, obtained divorces in more permissive states. But the Wright report indicated that only about 20 percent of divorces were obtained by couples who married in one state and divorced in another. He maintained that this was a reasonable figure for a nation with a highly mobile population, for many married couples moved from state to state. Wright speculated that as few as 3 to 10 percent of the 20 percent were divorce-seekers who temporarily migrated to other states to obtain divorces.[9]

On another issue, the report was alarmingly clear. Its tables and graphs dramatically illustrated the steady increase in numbers of divorces between 1867 and 1886. The report revealed, for example, that during 1879, the year that the *New York Tribune* had exhorted Americans to think about divorce reform, county courts granted 17,083 divorces in the United States. The agricultural state of Iowa, which was generally believed to be a family-oriented land of corn and contentment, accounted for 854. Only Ohio, Indiana, Illinois, and Michigan, all with considerably larger populations, exceeded Iowa in absolute numbers of divorces granted.[10] These findings confirmed what many Americans had feared—that divorce flourished even in conservative rural areas—and fueled the debate over divorce reform measures.

Dike was impressed by the report's thoroughness. He also felt that it gave him direction concerning divorce reform. In his 1889 report, he argued that because only a small percentage of divorces appeared to be migratory, the need for uniform divorce laws was less pressing than some advocates claimed. Although uniform divorce laws might lessen migratory divorces, these divorces were so few in number that the effect of uniform laws on the overall divorce rate would be minimal. "The centre of the problem of Divorce legislation is now proven to be located elsewhere than in the question of uniformity, though this is by no means eliminated from the problem." Dike recommended that states enact statutes requiring a waiting period between application for divorce and a divorce hearing, a mandatory waiting period before remarrying after a divorce, and reduction or abolition of "loosely defined causes of divorce."[11]

Like Dike, many members of Congress were also dubious about uniform divorce laws, but for different reasons. They believed that the Constitution would have to be amended before they could enact such legislation. An amendment of this nature would interfere with the privacy of the individual and the right of citizens to make decisions about their personal lives. Staunch defenders of states' rights also saw national divorce legislation as a threat to states' long-standing regulation of the marital status of their citizens. Still, beginning in 1884, at every session of Congress, members considered motions suggesting that the Constitution be amended. None garnered enough support to come to a vote.[12]

It appeared that uniform marriage and divorce laws would have to originate with the states and territories. Accordingly, in 1889 and 1890, Governor David B. Hill of New York recommended that the New York legislature create a "Commission for the Promotion of Uniform Legislation in the United States." The locus of discussion had shifted from New England to New York during the late 1880s because Hill and others hoped that uniform laws would prevent New Yorkers from fleeing the state's sole ground of divorce—adultery—and obtaining divorces in more lenient states. Apparently unimpressed by Wright's interpretation of numbers of migratory divorces, they wanted to stem what they believed was an exodus of unhappily married people.

In light of his previous position, it was surprising that Dike concurred with Hill's suggestion. Although he continued to lack enthusiasm for uniform divorce laws, Dike believed the proposal should get a fair hearing by representatives from all states and territories. In his 1891 report, Dike reminded League members that the American Bar Association had established a Committee on Uniform Legislation and that the governors of Massachusetts, New Jersey, Pennsylvania, Delaware, and Michigan had appointed state commissions to study uniform divorce legislation. He added that a number of state governors had responded favorably to the suggestion of the New York commission, the Bar Association, and the Divorce Reform League that a national conference on uniform divorce law be held.[13]

The creation of state commissions led to the formation of the National Conference of Commissioners on Uniform State Laws, which organized a meeting of state delegates in 1892. Although not all states joined, and disagreement among delegates blocked any notable achievements, Dike, whose organization changed its name to the National League for the Protection of the Family in 1897, continued to support the uniform law movement.

During the 1880s and 1890s, the movement gradually gained support. During these decades novelists increasingly commented upon social conditions and exposed ills of American society. Many actively called for a reduction in the divorce rate. They took the traditional view that marriage was a sacrament and a lifetime proposition to be terminated only for adultery and other weighty causes. Because they believed in the power and efficacy of legislation, some novelists enthusiastically supported uniform divorce laws.

Typically, these novelists' leading characters, usually women, clung to the sanctity of marriage. Even when their husbands divorced them and married other women, they refused to remarry until the death of the former husband set them free. The widely read and revered novelist E.D.E.N. Southworth especially supported the concept of marriage as a sacrament and lifetime agreement. Although she described marriage as a risky venture, Southworth never recommended divorce as an answer to its problems. The troubles of distraught spouses filled the pages of her books during the 1880s and 1890s, but the solution was always separation. This parting was often followed by reconciliation, or the death of one spouse, which freed the other from the marital contract.[14]

Other novelists focused on the destructive aspects of migratory divorce. An excellent example is Margaret Lee's little book *Divorce*, which was widely read during the 1880s and republished in 1889. In it, Lee proclaimed that the old-fashioned path of duty was best. In other words, only the death of one spouse released the other from his or her marriage vows.

In Lee's complex story, a woman of questionable morals named

Maude divorced her husband. Constance, clearly the heroine of the tale, criticized Maude's action, arguing that Maude could not remarry because her first marriage was sacred. Undaunted, Maude lured Constance's husband away from her. To this crass and selfish man, marriage was simply a contract to be terminated by divorce. "Expediency," he told Constance, was the "new morality." He continued, "this is the nineteenth century" and "there are modern remedies for the troubles that afflict human nature." Flaunting Constance's wish that they remain married, he left New York and obtained a migratory divorce in Connecticut. Even after her husband married Maude, however, Constance regarded her marriage as a lifetime agreement and refused to consider remarriage.[15]

William Dean Howells also presented the ugly side of migratory divorce in his widely read novel, *A Modern Instance,* published in 1886. It had first appeared in serial form in 1881, the year that the New England Divorce Reform Society was founded. In it, Howells described Marcia Gaylord's disastrous marriage to Bartley Hubbard and its traumatic end when Hubbard surreptitiously fled to Indiana to get a quick, uncontested divorce. Howell's tale, lauded as "rigidly moral" by a reviewer in the *Church Review,* vehemently condemned migratory divorce and the divergence of laws that made it possible.[16]

Still other novelists attacked permissive divorce laws. At the beginning of *Divorced,* published in 1887, Madeleine Dahlgren stated that her novel "was intended as a plea for the sacredness of the married tie, and also to exhibit some of the manifold dangers connected with our present system of divorce laws." To Dahlgren, human laws that allowed people to divorce contradicted divine laws concerning the sancity of marriage, undermined "the very foundations of the social structure," and fostered conspiracy, perjury, and licentiousness. After divorcing one wife and being divorced by another, Dahlgren's protagonist blamed his troubles on "this accursed facility of divorce" rather than on himself. He stated that he "so basely yielded to temptation" only because American laws failed to protect marriage as a "solemn religious rite." He added that if the United States, a Christian nation, enforced the marriage tie, he would be "a happy husband and a proud father." Instead, he had divorced; subsequently ruin befell him and he died of apoplexy.[17]

In 1894, novelist Margaret Deland took a similar stand in *Philip and His Wife.* After three years of tempestuous marriage, Philip came to believe that "marriage without love is as spiritually illegal as love without marriage is civilly illegal." He decided that he supported divorce, for "a priest's gibberish, a legal decree, or the tyranny of public opinion" was merely superficial bonds. When a friend tried to restrain Pihlip by arguing that a responsible person had to consider the good of society in such matters, Philip ignored him. The tale ended with Philip and his child living unhappily in one house, while his despondent former wife lived in another. The moral of Philip's story was that strict divorce laws were necessary to control "human selfishness."[18]

Besides novelists, many of the essayists who wrote in influential popular magazines of the late 1800s and early 1900s also supported the sanctity of marriage and disparaged permissive divorce laws. In 1880, for example, a writer in the *North American Review* complained that the idea of marriage as a sacrament and a lifetime agreement had "too little influence." He claimed that American individualism was the culprit, for individualism, "nothing more nor less than supreme selfishness," drove spouses apart. In his view, because the institution of the family undergirded the welfare of society, the state had to protect it with legal safeguards. In 1884, another essayist writing in the *North American Review* insisted that divorce was the foe of the American family, for it "invades the home" and "defiles its sanctities." He advocated government regulation and restriction of both marriage and divorce. And in 1893, an editorial in the *Nation* condemned South Dakota authorities for their negligent administration of divorce statutes, especially the loose application of grounds to divorce-seekers from other states.[19]

Other essayists bore out the worst fears of uniform divorce law advocates by presenting a dire portrayal of divorce. A 1902 article, for example, recounted wrenching case histories of children of divorce, all of whom suffered extensive trauma. Through no fault of their own, they were torn away from their mothers or fathers. Moreover, they lacked redress, rights, or claims to offset their terrible suffering.[20]

During the late 1800s and early 1900s, religious leaders also increasingly lent their support to the cause of uniform divorce law. Leaders of Protestant groups, especially the Episcopal church, joined with Roman Catholic leaders in speaking out against the growth of divorce in the United States. Episcopal Bishop William C. Doane of Albany, a strong supporter of New York's strict divorce law, spearheaded the organization of an Inter-Church Conference on Marriage and Divorce that met in 1903 and included representatives from approximately twenty-five religious denominations. In 1904, the *New York Times* reported that the American Baptist Home Mission Society had joined the growing list of religious groups favoring uniform divorce law. In 1905, the *New York Tribune* noted that Bishop Doane and a group of representatives from the Inter-Church Conference had urged President Theodore Roosevelt to lend his support to the cause of legislative change.[21]

President Roosevelt, who had been outspoken on the issues of marriage and divorce, was attuned to the mounting concern regarding divorce. Roosevelt was aware that Congress had been receiving numerous petitions requesting a second statistical investigation of marriage and divorce that would collect data from 1887, the year the first study ended, to the present.

When Doane and his committee called upon Roosevelt in late January 1905, to express their concerns about marriage and divorce, the President reassured them that "questions like the tariff and the currency are of literally no consequence whatsoever when compared with the vital question of having the unit of our social life, the home, preserved." He vigorously declared that "one of the most unpleasant and dangerous features of our American life is the diminishing birth rate, the loosening of the marital tie among the old native American families." Roosevelt then proclaimed that "no material prosperity, no business growth, no artistic or scientific development will count if the race commits suicide."[22]

At the committee's urging, Roosevelt issued a special message to the Senate and House alerting members that a growing number of Americans believed that the sanctity of marriage was held in "diminishing regard" because "the divorce laws are dangerously lax and indifferently administered in some of the States." He strongly recommended that they appropriate funds for an update of the Wright report. Members of both the Senate and House responded to Roosevelt's appeal by passing a bill authorizing the collection and publication of statistics on marriage and divorce between 1887 and 1906.[23]

As the overwhelming task of data collection began, Governor Samuel W. Pennypacker of Pennsylvania issued a call for a 1906 conference to be held in Washington, D.C.. The conference was to include representatives from as many states and territories as possible. Unlike Governor Hill's conferences, this would specifically focus on uniform divorce legislation. Pennypacker sent invitations to the governors of each state asking them to appoint delegates to a National Congress on Uniform Divorce Law to meet in Washington in February 1906.

The conference call alarmed women's rights leaders who had been following the progress of the uniform law movement with great interest and trepidation. They were distressed that women had been denied a role in shaping the divorce policies that could so deeply affect their lives. They were upset that men made the decisions about divorce statutes and procedures; men who, in their eyes, had failed to demonstrate any remarkable skills as leaders. Their view of male government was best summed up by Frances E. Willard's complaint to Susan B. Anthony that "men have made a dead failure of municipal government" and "an equally outrageous failure of Republic government."[24]

Elizabeth Cady Stanton was especially perturbed that men made and administered laws concerning marriage and divorce. According to her, men "have spoken in Scripture" and "in law." As husbands, men have decided the "time and cause for putting away" their wives. As judges and legislators, they had "entire control" over divorce proceedings. She preferred that woman, "the mother of the race," help make such decisions.[25]

Other women's rights leaders also believed that both women and men should be included in policy-making sessions regarding divorce. At the time that the first Wright report appeared, feminist orator and editor of the *Woman's Tribune,* Clara Colby, declared that "masculine legislation" regarding marriage underwrote the unequal position of women in marriage. According to Colby, the lowly status of wives created a "rush of women" to the divorce courts to obtain their "freedom from the galling bonds." There was no question in her mind that policy deliberations should include women.[26]

A number of women's rights leaders feared that male governors would fail to consider the appointment of female delegates to the conference on uniform divorce law. Accordingly, Alice Park, secretary of the California Equal Suffrage Association, wrote to the governors of every state, territory, and Alaska with a direct appeal: "We strongly desire that men and women shall be delegates to a national convention called to consider the relations of men, women, and children. The subject chosen for discussion is one of the most important subjects that requires attention, and the combined wisdom of the most carefully selected delegates of both sexes." Park added that in the forty-one states where women had "no political standing," the governors would perhaps resist the idea of appointing female delegates, but that the governors of the "free states," where women could vote, would surely think her request "timely."[27]

Park received eighteen responses from governors or their secretaries. The governor of Wyoming replied first, assuring Park that "some women" would be included in Wyoming's delegation. Others said that they had already appointed their delegates and one claimed to be unaware of the conference, but several promised that they would try to appoint a "lady" or two. The printed program of the conference reveals that, despite the governor's assurance, Wyoming sent only one delegate; the person's gender is unclear because initials, rather than a first name, were listed. Three of the "free states" sent women delegates: Idaho sent one; Utah, three; and Montana sent attorney and feminist Ella Knowles Haskell. The free state of Colorado sent one male delegate. Of the states that prohibited women from voting, Michigan and Washington were the only ones that sent female delegates. In total, forty-two states and territories sent representatives.

South Carolina, Mississippi, Kansas, Montana, and Nevada failed to send delegates. Because a number of states besides Wyoming listed only initials of delegates rather than first names, it is unclear precisely how many women and how many men attended.[28]

When the representatives convened in 1906, Pennsylvania delegates presented model divorce statutes for discussion. Most delegates realized that uniform divorce law would have to come from the states and territories rather than from the federal government because the movement to amend the Constitution so that Congress could pass family legislation lacked strong support. In addition, many delegates opposed the idea of such a constitutional amendment because they wanted their states to retain their long-standing right to control the marital status of their citizens.

As they examined the model statutes, the delegates discovered they agreed on a number of issues: a two-year residency requirement; personal notification of a defendant rather than notification by publication; public divorce hearings; and a one-year ban on remarriage. They sharply disagreed, however, on the idea of restricting grounds for divorce. New York's delegates especially opposed the long lists of grounds in other states. They feared that such extensive grounds would encourage New Yorkers to escape the sole ground of adultery in their own state by seeking divorce elsewhere. But delegates from several states balked at the idea of reducing the number of grounds offered to their citizens. After long and often bitter debate, a majority vote endorsed six grounds for divorce: adultery, bigamy, conviction for a felony, intolerable cruelty, wilful desertion for two years, and habitual drunkenness. In deference to strict states, notably New York, a strong disclaimer accompanied the list of grounds: "In those states where causes are restricted, no change is called for."[29]

The conference had degenerated into a battle between opposing factions. Perhaps, as one historian of divorce has speculated, those who wanted to restict divorce were clinging to the waning Victorian era and its emphasis on tradition, while those who accepted the expansion of divorce were looking to the emerging Progressive era with its focus on change.[30] Or, perhaps many Americans had adopted the viewpoint of a tongue-in-cheek editorial that appeared in the *Nation* in 1906. The author of "Divorce Too Difficult" labeled divorce procedures in the United States "backward." If people had the desire to part, and the money to pay for it, why should they be thwarted? The author also asked, "Why, in short, should the Constitutional right to the 'pursuit of happiness' be clouded and qualified by the scruples of backwoods legislators?" This author's solution to the rising divorce rate was permissive governmental regulation: "Let Congress provide a quick, quiet, and easy method of changing life partners."[31]

Dissension among conference delegates led to fulfillment of the *New York Tribune*'s prophecy that the conference would founder and fail to take any practical action. After a November follow-up conference that cast the conference's resolutions into model statutes, only New Jersey, Delaware, and Wisconsin adopted them. Leaders in all other states and territories found them unattractive for one reason or another.[32]

The response of women's rights leaders to the conference's lack of achievement was mixed. Some regretted its inconclusive outcome, for they believed in the cause of uniform divorce law. Belva Lockwood, for example, had made the issue of national uniform divorce law part of her platform when she ran for the presidency of the United States in 1884. Lockwood wanted wives to be equal with husbands in "authority and right" and hoped that uniform laws would help bring this about.[33]

Others were pleased that the movement for uniform law had collapsed, for they believed existing divorce laws offered protection to wives by allowing them to escape destructive marriages. They feared restricted divorce provisions designed largely by men would hurt wives by reducing the ease of divorce, thus forcing wives to remain in harmful marriages. Elizabeth Cady Stanton asserted on numerous occasions, "Liberal divorce laws for oppressed wives are what Canada was for Southern slaves." She feared uniform laws would be restrictive, thus nullifying existing divorce statutes that had been enacted "in a broader spirit." In her view, "legislative enactment" was simply a "fetish" of the day. The split between conference delegates made it "quite evident" that the nation was "not prepared for a national law." Because broad, pliable laws were Stanton's goal, she recommended the issue of divorce be left to state governments that were responsive to the "needs and convictions of the community."[34]

Still others women's rights leaders vehemently opposed all involvement with the issue of divorce and what they saw as its accompanying taint of free love. Frances Willard thought Susan B. Anthony's refusal to get involved in controversies and "theological fracases" was the right approach. In 1908, a young suffragist explained she wanted to keep the cause of suffrage pure, rather than identified with divorce and free love. In her view, women would get the vote only if they remained united in the suffrage cause rather than diffusing their energies in "tangential" issues.[35]

## Renewal of the Uniform Divorce Law Movement

The divorce controversy reached a new peak in public visibility when the government's second statistical study was released in 1908. This report on marriage and divorce between 1887 and 1906 described a far more comprehensive study of the topic than the earlier Wright survey. It contained statistics from 2,797 counties. Only six counties were omitted; for example, San Francisco County, California, was left out because an earthquake and fire had destroyed public records on April 18, 1906. This study asked for more information than the first, including data regarding alimony and number of children involved in divorces. In addition, it reflected a growing concern of the day, alcoholism, by compiling statistics of cases in which intemperance was a cause of divorce. Finally, it included a thorough digest of statutes concerning marriage and divorce in states, territories, and other countries.[36]

The second report shared one significant characteristic with the earlier Wright report, however. In stark tables and graphs, it demonstrated that the divorce rate was steadily climbing. The report stated, "movement, although occasionally checked or retarded by commercial crises, period of business depression, or other causes, has been almost without exception upward." In four years—1870, 1884, 1894, and 1902—the divorce rate was lower than in preceding years, but it exceeded previous rates in the twenty-nine other years studied.[37]

These figures energized the advocates of uniform divorce law and they quickly renewed their crusade on behalf of nationwide divorce statutes. In 1909, C. LaRue Munson, Chancellor of the Episcopal Diocese of Harrisburg and vice president of the National Congress on Uniform Divorce Law, proclaimed that three hundred years of the American "spirit of individualism" had created a chaotic situation. In other words, each state had enacted its own laws, too many of which were highly permissive. Adopting the relatively strict 1906 model statute was the only remedy. To Munson, "lax divorce laws" were "a menace to family life and all that makes it sacred."[38]

In the same year, Jane Burr's *Letters of a Dakota Divorcee* disclosed the ills of migratory divorce. By describing what was supposedly her own experience as a divorce-seeker in Sioux Falls, she confirmed public images of boardinghouse life in a western divorce colony, corrupt attorneys and paid witnesses, the haste with which divorcing spouses crassly chose new partners, and the casualness and outright perjury involved in divorce proceedings.[39]

The reawakened interest in national uniform divorce law continued for several years. In 1911, a lengthy *New York Herald* article listed the grounds for divorce in all states and territories. Its author argued that such diversity demanded revision; the "hodge-podge" of divorce laws should finally be brought into a harmonious whole. The following year, Governor James H. Hawley of Idaho urged participants in the annual conference of state governors to adopt the 1906 model statutes.[40]

Despite this brief regeneration, the uniform divorce law movement gradually lost momentum. It was increasingly apparent that states and territories could not, or would not, agree on divorce provisions. Around 1910, many supporters of uniform divorce law began to recognize the odds against their cause. The divorce rate was rising, divorce was gaining acceptance, and the National League for the Protection of the Family was waning in membership and visibility. It no longer provided impetus for the uniform law movement.

Still, some tenacious people clung to the idea of uniform divorce law. To them, the failure of states and territories to agree on this matter demonstrated that the federal government would have to mandate such laws. If a constitutional amendment was necessary, then they would campaign for such an amendment. President Roosevelt's annual message of 1906 had paved the way. Arguing that "the home life of the average citizen" was imperiled by the scandals and abuses resulting from wide variations in divorce laws, Roosevelt suggested "the whole question of marriage and divorce should be relegated to the authority of the National Congress."[41]

Boards of Episcopal, Methodist, and Roman Catholic clergy, several members of the U.S. Congress, the General Federation of Women's Clubs, and the California Commission on Marriage and Divorce also supported the idea of a constitutional amendment. In 1913, Charles Franklin Thwing, a Congregational minister and president of Western Reserve University in Cleveland, joined the fray. Much like Munson, he railed against American individualism and recommended a constitutional amendment vesting in Congress the power to regulate divorce. These uniform law proponents and their successors continued to unsuccessfully push the idea of a constitutional amendment until 1947.[42]

Were the uniform divorce law proponents total failures? Ostensibly yes, because their restrictive provisions were rejected. Thus, the divorce rate continued to rise and western divorce mills continued to function. Still, uniform divorce law advocates were successful in a sense, for they forced many Americans to explore and confront the issue of divorce. In addition, leaders of the uniform divorce law campaign convinced reformers, legislators, policy-makers, and individual citizens that it was necessary to collect statistical data before making generalizations about divorce or undertaking corrective measures.

Why were they unable to achieve more than this? The uniform divorce law movement failed to attract a sufficient number of supporters because of a combination of internal mistakes and uncontrollable outside forces.

Specifically, the leaders of the uniform law movement committed an error when they neglected to include more women in policy-level conferences. This inattention to women, especially feminists, turned a visible, vocal group against their cause. As early as 1890, in the pages of the widely read *North American Review,* Mary Livermore attacked divorce laws made by men to serve the interests of men. She demanded that women be included in policy deliberations.[43]

The leaders of the uniform divorce law campaign also failed to construct a broad platform that would attract people who advocated their own solutions. Livermore, for example, insisted that "legal equality" between mates was the only solution to growing "restlessness and unhappiness in married life." And an outspoken woman essayist, Marguerite Wilkinson, argued that only education could cure the "divorce evil." The "right education" would help people develop informed minds and strong characters, crucial qualities in marriage partners. Wilkinson also believed people must be taught about sex and love before marrying rather than being "childishly ashamed" to talk about such matters. Like Livermore, Wilkinson argued for equality of wives and husbands based upon a partnership of "responsibilities, duties and powers." If young women and men studied, played, and worked together, they would get to know each other rather than making their choice of spouses by "proximity or moonlit perfervid error."[44]

Uniform divorce law proposals were narrow in another way: they failed to raise such crucial issues as alimony and child custody and focused on reducing the divorce rate and destroying western divorce mills. Although Mary Livermore objected that a mother often lost custody of "little children, whom she wins in the valley of death, at the risk of her own life," uniform law advocates failed to discuss this and related issues.[45] In so doing, they failed to attract the support of people whose major concerns about divorce were alimony, child custody, and easing the burdens on divorcing spouses and their children. Uniform divorce advocates seemed to overlook the desirability of revising laws for the benefit of divorcing Americans.

External forces also impeded the success of the uniform divorce law campaign. One influential force was social scientists' views of divorce, views that opposed the idea of uniform divorce law at almost every turn. The other was a rising divorce rate; even as some Americans harangued against divorce, other Americans increasingly embraced it.

During the closing decades of the nineteenth century and the opening decade of the twentieth, social scientists tended to argue that divorce was a positive institution. Many believed that it especially benefited modern society by eliminating dysfunctional marriages. This view began with Herbert Spencer's *Principles of Sociology,* published in 1876, and Lester Ward's *Dynamic Sociology,* published in 1883, both of which endorsed easy divorce.[46]

Later social scientists used Census Bureau statistics to develop an additional contention: divorce statutes had very little effect on the divorce rate. Therefore, legislative changes—whether uniform law or constitutional amendment—would have minimal impact on the number of divorces granted. Walter Willcox, a statistician at Cornell University, unequivocally stated in 1891 that legislation had little or nothing to do with the divorce rate: the effect of divorce laws was "subsidiary, unimportant, almost imperceptible." To Willcox, the significant factors pushing the divorce rate upward were industrialization, the emancipation of women, and the "spread of discontent." Willcox argued that a solid education in moral values, especially classroom lessons on the "relations and duties of home life," would offset these factors and keep marriages together.[47]

George E. Howard, eminent historian and sociologist at the University of Chicago, agreed with Willcox that divorce statutes failed to affect the divorce rate in any significant way. Writing in 1904, Howard theorized that restrictive divorce statutes in one state did little to reduce the total number of divorces; they simply caused petitioners to seek divorces in other jurisdictions. Although Howard thought that "simple, certain, and uniform" divorce legislation would eliminate lax interpretation and application, he believed that it would fail to reduce numbers of divorces. In his view, divorce stemmed from certain fundamental causes, especially "false sentiments" regarding marriage and the family. Therefore, divorce was "a remedy," not a "disease." In other words, divorce was simply the legal process that released spouses from an unworkable marriage. Because disintegration had begun long before the couple reached the divorce court, Howard believed improving marriage relations should be the focus of reform and policy-making.[48]

In 1908, the third annual meeting of the American Sociological Society focused on the topic of the American family. The association's president, William G. Sumner, opened the conference by stating that the family "has lost its position as a conservative institution and has become a field for social change." A number of scholarly papers and Charlotte Perkins Gilman's summary of her influential book *Women and Economics* reiterated the damage sustained by the family and indicted as the culprits the twin forces of industrialization and urbanization. Several sociologists, notably George Howard, Edward Alsworth Ross, and James Lichtenberger, added that increasingly free divorce was a positive development in this situation, for it terminated marriage problems created by industrialization, urbanization, and other factors, including Howard's "false sentiments."[49]

The following year, Ross expanded this point of view. Although restrictive divorce statutes failed to reduce the divorce rate, the family would endure: "the failing grip of the legal institution" did not necessarily mean "a corresponding abandonment of the hallowed ideal of marriage as a lifelong union." Divorce neither destroyed marriages nor tarnished the purity of home and family. "Evidently," he commented, "there is a widespread failure to distinguish between symptom and disease." Ross agreed with Munson and Thwing that American individualism was related to the rising divorce rate. Long-term unions were impossible in a country where men and women worshiped individualism. In addition, industrialization weaned women away from the domestic arts and offered them a way to support themselves after divorce. Ross's solution to marital disintegration was a thorough education of American youth in "the ethics and ideals of the family," accompanied by stricter marriage requirements.[50]

In 1909, James P. Lichtenberger's *Divorce: A Study in Social Causation* cogently summarized the growing chasm between uniform law advocates and social scientists. He noted that "temperament and training" influenced the way an observer saw the rising divorce rate. One group—reformers, moralists, and religious dogmatists—interpreted the growth of divorce as a sign of social disintegration and breakdown of the family. The other group—including sociologists—saw the rising divorce rate as the result of environmental changes: a societal adjustment to modern civilization, especially industrialization and urbanization. According to Lichtenberger, the latter interpretation was propounded by fair, objective social scientsts. Their interest was only a "scientific investigation of facts." Lichtenberger concluded that although the divorce rate would continue to rise, the family would survive.[51]

By 1910, social science had become so widely accepted, and even idealized, that the stance of social scientist did a great deal to erode the arguments of uniform divorce law advocates. Instead of supporting legal changes, most social scientists disparaged their effects. Rather, they focused on more concrete measurable factors, especially industrialization and urbanization as causes of the rising divorce rate. They looked to expected changes in marriage to prevent divorce rather than to changes in divorce statutes.

The rising divorce rate was the second external factor that defeated the national uniform divorce law movement. Many Americans began to accept the ubiquity of divorce in American society; even as uniform divorce law advocates called for a reduction in divorces, increasing numbers of American divorced. Because thousands of people each year rejected the argument that marriage was a sacrament and a lifetime agreement, and discounted warnings regarding the decline of American society and their children's futures, the divorce rate continued its upward spiral in the United States.

In 1906, the same year that the Congress on Uniform Divorce

Laws met in Washington, D. C., the Massachusetts Bureau of the Census released the following figures regarding divorce in that state for the years 1860–1904. Total divorces 31,553; women 21,685 (69.3%); men 9,688 (30.7%). These figures indicated that in 1860 there was one divorce to 51.05 marriages contracted; by 1904, this figure had jumped to one divorce to 15.31 marriages.[52]

When the findings of the Census Bureau's second statistical survey of marriage and divorce in the United States for the years 1887–1906 were published in 1909, they revealed a similar rate of growth in divorce on the national level. Total divorces 945,925; women 629,476 (66.6%); men 316,149 (33.3%). The number of divorces nationwide increased at a ratio of approximately 30 percent every five years. Numbers of divorces also grew about two and one-half times as fast as the general population and the number of marriages. The nation's divorce rate was one divorce to every 13.9 to 15.6 marriages, depending upon how the statistics were interpreted. The report concluded that the United States divorce rate exhibited a "marked persistency" in increase that was "almost without exception upward."[53]

This second study disclosed other significant aspects of the expanding institution of American divorce. It reported which grounds were the most popular (see Table, below). The dominant trend was a move from heinous grounds, notably adultery, to less traumatic charges, especially cruelty.[54]

The tendency to utilize less offensive and perhaps less accurate grounds reminds us once again to read divorce petitions with caution, for they were not always literal, complete descriptions of the reasons for a marriage's disintegration. A wife who charged cruelty rather than adultery may have been protecting herself from scandal rather than reporting the actual causes that drove her to seek a divorce.

In addition to grounds, the Census Bureau also determined the occupations of about 75 percent of the husbands involved in divorce cases. These data indicated that although people of all occupations

### United States, 1887–1906

|  | Total Divorces | | Women's Divorces | | Men's Divorces | |
|---|---|---|---|---|---|---|
| adultery | 153,759 | 16.3% | 62,869 | 10.0% | 90,890 | 28.7% |
| cruelty | 206,225 | 21.8% | 173,047 | 27.5% | 33,178 | 10.5% |
| desertion | 367,502 | 38.9% | 211,219 | 33.6% | 156,283 | 49.4% |
| drunkenness | 36,516 | 3.9 | 33,080 | 5.3% | 3,436 | 1.1% |
| neglect to provide | 34,670 | 3.7 | 34,664 | 5.5% | 6 | * |

*less than one-tenth of 1 percent

and social classes divorced, divorce was more prevalent among the lower classes. Actors and professional showmen exhibited the highest rate of divorce, a phenomenon that presaged the entertainment industry's high divorce rate during the twentieth century. Musicians and teachers of music ranked second, while commercial travelers were in third place. Farmers and clergymen were at the bottom of the list. Although the study neglected to determine the occupations of wives, individual cases reveal that divorcing women held a wide variety of jobs, including various types of domestic service and professions including newspaper reporter and doctor.[55]

Ethnic backgrounds and religious affiliations of litigants went unmentioned in the government study because local divorce records omitted such information. Anecdotal evidence, however, suggests that members of such tradition-oriented groups as immigrants and Jews were not divorce-free. An 1888 divorce case in San Diego, for example, involved a Jewish couple who had married in their native Poland in 1863. After David Marks gambled away their funds and deserted her on several occasions, Fannie Marks, a successful boardinghouse proprietor and mother of thirteen children, filed for a divorce in San Diego in 1888. The case ended when David shot her in the leg and himself in the head.[56]

The Census Bureau report did, however, analyze regional variations in divorce. Although all regions of the United States were experiencing rising divorce rates, the western states and territories had the highest rate, the northeastern states had the second highest, and the southern states the lowest. Given the strength of the Anglican church, the perseverance of traditional attitudes toward women, restricted economic opportunities for women, and a high proportion of rural inhabitants in the South, it is not surprising that the South had a lower divorce rate than northeastern or western areas.

In addition, although judicial decisions sometimes softened their impact in practice, restrictive divorce statutes remained in force in many southern states. Alabama statutes, for instance, continued to allow divorces of bed and board long after many states had judged them to be ineffective and inhumane. Alabama divorce statutes also restricted the charge of cruelty to physical violence and allowed chancellors to prohibit guilty parties from remarrying. In Louisiana, a progressive 1898 statute allowed divorces of bed and board to become absolute divorces if one spouse or the other made application, but fathers routinely received custody of children unless there was "sufficient" reason to deprive them of it. Louisiana statutes also denied the guilty party in an adultery suit the right to marry subsequently his or her "accomplice," and prohibited a divorced wife from remarrying for ten months after a divorce, presumably to allow any children of the first marriage to be born before she contracted another union.[57]

In other southern states, divorce statutes were similarly restrictive. Maryland law allowed divorces of bed and board, lacked a cruelty clause, and restrained guilty parties in adultery cases from remarrying during the lifetime of the former spouse. Mississippi statutes defined cruelty as physical violence, lacked provision for temporary alimony, and prohibited an adulterer from remarrying. Virginia statutes allowed divorces of bed and board, lacked a cruelty clause, and gave chancellors the power to prohibit a guilty party from remarrying, although case records indicate that chancellors seldom did so. In 1881, for example, Mildred F. Gibson obtained a divorce in which both parties were permitted to remarry, even though her husband had been judged guilty of the serious crime of adultery.[58]

Between 1887 and 1905, North Carolina legislators expanded the state's statues, then retracted the new provisions. In 1887, lawmakers added the grounds of a husband's conviction of a felony and a husband's absence for one year to the existing grounds of adultery, impotency, and a wife's pre-existing pregnancy. In 1888, they added a wife's refusing to have "sexual intercourse" with her husband for twelve months. In 1895, they recognized abandonment for two years as a ground. In 1899, they added a husband's taking his wife to another state and mistreating her as a ground for divorce. But on March 6, 1905, the legislature cut back these provisions; grounds for divorce were again only adultery, impotency, and a wife's pregnancy.[59]

Commissioner of Labor Carroll D. Wright noted that, despite these statutes, the southern divorce rate had risen somewhat. He speculated that the southern divorce rate was rising because a growing number of African Americans were seeking divorces from courts rather than their own churches and communities. Although Wright's field agents had tried to determine the race of litigants, they were stymied because race was seldom noted in court records. They did locate a number of court officials and southern divorce attorneys who estimated that as many as 50 to 90 percent of divorces went to African Americans. In addition, field agents identified a substantial number of divorced African Americans in southern states in 1900. Still, no clear pattern of African American divorce emerged from the statistics. Wright observed that "it seems impossible to draw any definite conclusion from the figures" and to establish "any definite fact in regard to the comparative prevalence of divorce" between blacks and whites. Wright strongly recommended that in the future divorce documents include race of litigants.[60]

As revealing as these statistics were, they provided only one measure of marital breakdown in the United States; omitted were marriages that ended in separation and desertion because of the lack of records documenting these occurrences. Yet individual cases attest to the continuing existence of desertion in the United States. One case was that of teacher and African-American activist Mary McLeod Bethune who married Albertus Bethune in 1898 at Sumter, South Carolina. She bore one child, Albert McLeod Bethune, in February 1899. Ten years later, she left her husband, who she said was devoid of any burning ambitions of his own and often expressed doubts about her career. Although the Bethunes never became a divorce statistic, they represented untold numbers of dissatisfied spouses who parted without benefit of a legal decree.[61]

## Self-Help Literature

As the divorce rate rose in the United States, self-help authors increasingly tried to assist their readers in achieving satisfactory marriages rather than resorting to divorce. Although it is impossible to know how many people bought and read advice manuals, such books appeared on the market with regularity. This suggests that self-help books had a sizable audience, for publishers surely would have abandoned the genre if it failed to sell.

Self-help authors tried to combat the rising divorce rate in their own way, by offering practical counsel on how to avoid divorce. They attempted to strengthen marriages and avert people from the divorce courts by suggesting ways of choosing a good marriage partner and achieving serenity and happiness in marriage.

An early example was Hiram Pomeroy's *Ethics of Marriage,* published in 1888. Pomeroy, a medical doctor, believed that hasty and "spurious" marriages too often led to "infelicity," divorce, or the murder of one spouse by the other. He cautioned his readers to marry for love rather than wealth or status. Women had to be especially careful because they were choosing husbands *and* fathers for their children. To aid their decisions, women had to assess realistically both a man's positive and negative qualities. "It is a foolhardy thing for a woman to marry a man to save him," Pomeroy warned. She would be better off as a "spinster" then entering "dishonorable wifehood."[62]

In 1892, popular author Eliza Chester offered her readers similar advice. She emphasized that successful marriages were based upon love, respect, and "sympathy of tastes." According to Chester, women who failed to find such a match often chose to remain single. According to Chester, single women constituted a "great army;" they all believed in marriage, but only in happy ones.[63]

Two years later, medical doctor B. G. Jefferis inveighed against loving people "because they have money." He counseled people to court carefully if they hoped to avoid joining the ranks of wretched, mismated spouses. A wise woman would give herself time to find a man who treated his mother and sisters well and was kindhearted, generous, and unselfish. Above all, she would avoid marrying "an intemperate man with a view of reforming him." A wise man would look for a woman with modesty, religious sensibilities, purity, and a willingness to work hard. Choose wisely and well first, Jefferis cautioned, and then fall in love.[64]

In 1899, author Charles Sargent took yet another tack. He argued that the growth of divorce stemmed from a loss of privacy in marriage. A married couple must keep their secrets to themselves, thus building confidence and trust in each other. In addition, each partner should have his or her own physical space: "a room sacred from intrusion." Sargent hastened to add that he was not calling for reserve, indifference, or estrangement. On the contrary, he strongly recommended mutual decision-making and regular communication between spouses.[65]

In 1903, another medical doctor, Mary R. Melendy, declared that marriage could be sacred and joyous; its misery came from "wrong, foolish" choices. Partners must be "congenial" in "spirit." And men must be willing to control their sexual appetites, for conceiving children when "any degree of reluctance or want of preparation exists on the part of the mother" was unconscionable.[66]

In 1909, writer Anna B. Rogers insisted that shaky marriages were founded on "the latter-day cult of individualism; the worship of the brazen calf of Self." She especially blamed divorce on those women who refused to recognize that marriage was "their work." Women had lost the art of "giving," Rogers added. Instead, they had become selfish and individualistic, carrying "the germ of divorce" in their "veins."[67]

Each of these writers tried to stem the tide of divorce by suggesting ways that marriage could be improved. Their advice focused on strengthening marriage rather than on assisting divorce-seekers. Few writers addressed such matters as locating an attorney, laws of various jurisdictions, child custody, property settlements, alimony, and general etiquette of divorce. One of the few practical guides to appear was *How to Get a Divorce*, published in 1859 in New York City. Written by an attorney, this small pamphlet contained a compilation of each state's divorce laws, but neglected to give step-by-step advice.[68]

During the late nineteenth and early twentieth century, many Americans clung to traditional views of marriage and divorce. They failed to realize that powerful forces, including industrialization, urbanization, changing gender roles, and rising expectations of marriage, propelled people toward divorce. Instead, they believed that strict divorce statutes could keep people in their marriages, reduce the divorce rate, and eliminate western divorce mills.

Other Americans, however, acknowledged divorce as a reality. Increasingly, advocates of divorce argued that divorce was a citizen's right, and that it was beneficial to American society because it eliminated dysfunctional marriages. Although "no thinking person" would argue that divorces "be granted people who have simply tired of the marriage yoke," certainly divorces for "extreme cruelty, drunkenness and certain forms of crime" were necessary. Should divorce disappear, "women would be the chief sufferers, for they would be compelled to bow their necks to the yoke." Rather than escaping unbearable marriages, aggrieved wives would have to endure—to be "a bond slave as abject as she was in the days of the savagery of the race."[69]

Shortly after Samuel Dike died in 1913, the League, now largely a one-person operation, died as well. A small group of people would continue to push the issue of uniform divorce laws briefly into the public spotlight, but divorce was achieving a legitimacy in American life that would defeat their efforts. Those who believed that "truly the land needs a reform" had waged a fierce battle, but the growing commitment to the right of individuals to solve martial problems by way of the divorce court had overwhelmed, and would continue to overwhelm, their cause.

# *Asunder and Adrift in Early Twentieth-Century America*

☐ On May 18, 1913, Sara Bard Field boarded a ship in Portland, Oregon. With her sister Mary and her four-year-old daughter Kay, Field sailed to San Francisco where she spent $62.70 on three train tickets to Goldfield, Nevada. During the train's Sacramento stop, she mailed a letter to her husband telling him of her plans. When the trio arrived in Goldfield, they discovered what Field described as a "typical mining town," but to their amazement and delight, Goldfield's hotel boasted running water, bathtubs, electricity, and "excellent food." After settling into the hotel, Field met with an attorney to initiate a divorce suit against her husband, Albert Ehrgott.[1]

Sara Field was typical of a growing number of Americans who rejected the idea of marriage as a sacrament and for life. Field was willing to put her marriage asunder and set herself and her children adrift because her watchword was love. Although her husband had not committed a marital crime—adultery, desertion, non-support, or cruelty—Field no longer loved him. Moreover, she loved another man. Thus, she was willing to thwart her husband's opposition to a divorce by seeking a migratory decree in Nevada.

Many Americans still opposed this kind of thinking. As a result, the great American divorce debate continued. Most of its themes were familiar ones. Was marriage a lifetime undertaking or was it a dissolvable contract? Should divorces be difficult or easy to obtain? Could uniform national divorce laws reduce the rising divorce rate? What actions would curb migratory divorces? What factors caused the rising divorce rate?

Certainly, a new morality seemed to be developing in the United States. As the first decade of the twentieth century came to a close, divorce appeared to be everywhere; it also seemed to have garnered widespread support. Progressives had joined with feminists and oth-

130

*Godey's Lady's Book* presented a romanticized view of love. *Godey's* 49, 2 (February 1850).

*Peterson's Magazine* presented a similarly romanticized view of courtship. *Peterson's* 55, 1 (January 1860).

Newspaper editor Horace Greeley was a frequent and bitter critic of divorce during the mid-nineteenth century. Courtesy of the National Archives, Washington, D.C.

Woman's rights leader Elizabeth Cady Stanton was an articulate supporter of divorce during the mid- and late nineteenth century. Courtesy of the National Archives, Washington, D.C.

The *Police Gazette* was especially critical of the Oneida Community of New York. It alleged "obscene orgies and pernicious teachings" among the "saintly sect." Courtesy of Estelle B. Freedman, Stanford University.

Well-known minister Henry Ward Beecher found himself embroiled in the Tilton divorce scandal during the early 1870s. Courtesy of the National Archives, Washington, D.C.

Mormon leader Brigham Young's practice of granting church divorces to polygamous couples between 1847 and 1876 caused many people to view Utah as a divorce mill. Courtesy of the National Archives, Washington, D.C.

# $50 REWARD!

Reward of $50.00 will be paid by Zepheniah Smith for the recovery or information leading to the recovery of Mrs. Nannie Bell Smith; age 26 years, height about 5 feet 4 inches, weight 120 pounds, brown eyes, dark brown hair. Disappeared from her home near Trousdale, Okla., Oct., 20, 1905. Was traced to Oklahoma City, Okla. Will likely seek employment such as light house work, as health would not permit her to do hard work. Was in very poor health at time of disappearance. The above picture is good of her taken about a year ago.

## ADDRESS ALL INFORMATION TO
## ZEPHENIAH SMITH
### Trousdale, Okla.

Box No. 3.

"Wanted" Poster, Oklahoma Territory, 1905. Courtesy of the Oklahoma Territorial Museum, Guthrie.

# DISAPPEARED.

FRANK LIMBACK of Owensboro, Ky.; gone since October 11, 1906. He is about 30 years of age; 5 feet 10 inches high; weighs about 175 pounds; smooth shaven; heavy cheek bones; gray eyes; Roman nose; rather sharp chin; heavy deep set jaws; powerful muscular build. Two years ago last September he received serious injury to his spine in a railroad wreck at Duncan, Indiana. Married, has wife and two children. By profession a blacksmith and a railroad man. He left Owensboro for Jasper, Indiana, Oct. 11th, never heard of since. Walks on crutches. Wife and children in destitute circumstances. Any information to the undersigned will be greatly appreciated.

## MRS. FRANK LIMBACK,
### OWENSBORO, KY.

"Wanted" Poster, Kentucky, 1906. Courtesy of the Oklahoma Territorial Museum, Guthrie.

Minnehaha County Courthouse in Sioux Falls, South Dakota, where divorce suits were heard during the 1890s. Courtesy of Glenda Riley.

One of many beautiful homes built in Sioux Falls during the 1890s. Critics charges that such mansions were built with proceeds from the town's divorce industry. Courtesy of Glenda Riley.

One of many Las Vegas wedding chapels that accommodate newly divorced people who wish to marry a new spouse immediately. Courtesy of Glenda Riley.

Despite the rising divorce rate and the divorce mill panic of the 1890s, Americans continued to believe in marriage. A courting couple near Sioux City, Iowa, *ca.* 1900. Courtesy of Glenda Riley

Sara Bard Field and Charles Erskine Scott Wood in 1943 shortly before his death. Field obtained a migratory divorce in Nevada to live with Wood. Courtesy of the Henry E. Huntington Library.

ers in calling for the freedom of choice to stay in a marriage or leave it.[2] And in 1910, a former justice of the Supreme Court expressed an increasingly prevalent attitude when he stated that he failed to understand why the partnership created by marriage differed from a commercial partnership; why the first could be dissolved "while the other is indissoluble."[3]

The following year, novelist and radical love theorist Upton Sinclair found himself embroiled in a scandalous situation that put divorce on the front page of American newspapers for days at a time. When his wife Meta became involved with poet Harry Kemp, Sinclair reacted in a traditional way by suing her for divorce and naming her lover as a co-respondent. Meta explained she was doing what Sinclair had preached; she was seeking her independence and self-fulfillment. Her quest, journalists revealed to an entranced public, supposedly involved a sexually vital man as opposed to her sexually inadequate husband. After several divorce hearings in New York courts made it clear that the state's restrictive divorce statutes would prevent Sinclair from getting a divorce, he traveled to the Netherlands, where he obtained a migratory divorce.[4]

Emma Goldman also regularly pushed the issue of divorce into the public spotlight. After she fled from a brief marriage undertaken when she was seventeen, she began to argue her unorthodox views on the speaking platform, from behind jail bars, and in the pages of the radical *Mother Earth News,* which she founded and edited between 1906 and 1918. Red Emma, as she was known, insisted that man-woman relationships must be based on equality for "a true relation of the sexes will not admit of conqueror and conquered." She supported free love, and she advocated birth control six years before Margaret Sanger began her campaign. As an editor, Goldman frequently published the work of authors who called for complete freedom of divorce, branded marriage obsolete, and espoused total sexual freedom.[5]

But the new morality was not adopted just by radicals and various types of reformers. Sara Field was a typical, middle-class American, yet the moral stance that underwrote her divorce and subsequent actions would have been incomprehensible to earlier generations of Americans. Field explained that her divorce from Albert Ehrgott was inevitable because she was only a "girl" of eighteen when she married him. As she matured, she grew away from him. And as she became interested in such causes as women suffrage, he became more "narrow and inflexible . . . fanatical and intolerant." After "much physical, mental, and spiritual suffering," she had decided on divorce by early 1913, when her doctor sent her to a tuberculosis sanitarium in Pasadena, California.[6]

Faced by her husband's opposition to a divorce, Field resolved to go to Nevada to obtain a divorce, preferably to a town that was far from "the unpleasant notoriety of Reno." When Field met with her attorney in Goldfield on May 23 of that year, he told her he could get her a divorce decree in less than the required six months if her husband would give his consent. Although Field was euphoric—"what a wonderful relief it would be if I could have this case come up at once and not have to endure this dreary six months waiting"—Albert continued to oppose her. Field rented for $20 a month a small four-room house, where she and her daughter Kay lived until Field's divorce was granted in early 1914, after what she termed "long, unnecessary delays."[7]

Albert Ehrgott, a Baptist minister, described the break-up of his marriage to Sara differently. Ehrgott blamed his wife's alienation from him on attorney, writer, and reformer Charles Erskine Scott Wood, whom Sara had met in 1910 when she was twenty-eight and Wood was fifty-eight. Ehrgott wrote Wood accusing him of corrupting Sara with "free love poison" and warned Wood to stay away from his wife. Ehrgott also wrote to Sara's sister Mary, saying Sara had been happy with him until "Mr. Wood came along." He added that "certain unsurmountable reasons make it absolutely impossible to grant Sara any legal release from her marriage vows."[8]

Divorce Albert she did, however. Nearly three years later, Charles Wood retired from his law practice and left his wife, a staunch Catholic who refused to divorce him. He and Sara established a home together in San Francisco, an action that elicited a range of opinions from Sara's family. Sara's eldest child, Albert, who was five years older than her daughter Kay, wrote to Wood: "I hope that it won't be so very long now when people who love each other will not have to have a few silly damn fool words said over them by a parson to make them a so-called husband and wife." Albert added that he had never seen "a love so great and wonderful and beautiful" as that shared by his mother and Wood.[9]

Ehrgott also wrote to Wood, but in a very different tone. He told Wood that his "dribbling philosophy of free love is but a futile excuse for an unholy indulgence of 'love' and a sacrilegious intrusion into another man's home which an aroused public conscience will not much longer tolerate." Ehrgott accused Wood of stealing another man's wife, trifling with God, and defying public opinion. He refused to let Sara have any time with Kay, and when young Albert died, he blamed Sara and Wood for the tragedy.[10]

Wood had little use for Erghott, but in 1927, he put his philosophy of marriage into words in an attempt to reconcile his estranged son, Erskine, to his relationship with Field. Wood optimistically wrote of his belief that "old, narrow and early crude conventions" based on the "right of possession and the justness of jealousy" were breaking down; "conventions founded on the right of every individual to his own soul and his own life" were gradually replacing the older notions. Because so many marriages ended in "shipwrecks," the "old idea of the sacrament of marriage" was fading as well. If marriage was based upon the "mutual love of the parties and the mutual desire for companionship and mutual willingness to live together," then, "when this mutual desire to live together ends," the marriage relation also ends.[11]

To Wood, it was unethical to force two people to live together "against the will of either." He was baffled that so many people failed to see that free divorce was "best for society and the race"; every couple forced to remain together despite altered feelings was "a cancer" and "every refusal of freedom leads to falsities." Wood swore that he and Field would never marry; they would reject "the archaic superstition and falsity" known as marriage.[12]

The new morality that affected a wide spectrum of Americans was accompanied by a rising divorce rate. Regardless of how the divorce rate was computed, its overall pattern was the same: upwards. Whether the number of divorces between 1910 and the mid-1940s was compared with total population, married population, or marriages in a given year, the resulting ratio showed an ascending curve that surpassed the divorce rate in all other nations. The divorce rate was abnormally high after World War I, when hasty wartime marriages collapsed and others succumbed to war-related stresses, but it leveled off somewhat during the 1930s Depression when limited resources mitigated against divorce. In 1928, a year approximately midway between 1910 and 1945, slightly more than one of six marriages ended in divorce.[13]

Most other trends held steady as well. Urban divorce rates continued to out-distance rural rates. The West's divorce rate continued to exceed that of other regions. Women continued to obtain more divorces than men; in 1928, women received 71 percent of all divorces granted. And in that year, 47 percent of divorces were granted on the increasingly popular ground of cruelty, a charge women continued to use more than men.[14]

As more sophisticated data became available, analysts confirmed some suspected trends, especially that divorce was higher among African Americans than among whites. Interviews with poor, southern blacks conducted by Works Progress Administration workers during the 1930s indicated that African-American marriages were bedeviled by problems of adultery, abuse, desertion, and general marital misery. In the view of one justice of the peace, the United States needed easy divorce "for the sake of the pore folks" who often suffered marital ills.[15]

Despite evidence of separation and divorce among African Americans, it was impossible for analysts to determine whether race was the crucial variable in the African American divorce rate. Perhaps the African-American divorce rate was high because African Americans tended to be poor, uneducated, and unemployed—all significant factors in marital breakdown. High mobility among African Americans and racial discrimination against them may have also adversely affected their marital stability.

It was also impossible to determine the full extent of marital breakdown in the United States because divorce was only one way marriages ended. Separation and desertion continued to be two other widely used alternatives. Some separated couples and deserted spouses were too poor to obtain divorces; others feared social stigma, or dealing with financial or family complications. Although no records document separation and desertion, a study of Philadelphia in 1922 and of Chicago in 1927 indicated that desertion was especially high among immigrant groups that had relatively low divorce rates. In addition, a 1928 survey of community organizations revealed that 134 out of 145 respondents thought desertion was a vexing problem in their communities. Still other people terminated their marriages through annulments, which were sometimes included in divorce statistics but other times went uncounted.[16]

Leaders of the uniform divorce law movement, which experienced something of a resurgence during the 1920s, maintained that they had a solution to the rising divorce rate. Supported by the powerful General Federation of Women's Clubs, they argued for the passage of a constitutional amendment that would give Congress the power to establish uniform divorce statutes in all states and territories. Such statutes would, they believed, prevent hasty marriages, make divorce more difficult, and eradicate migratory divorce.[17]

At the urging of Federation officers, on January 11, 1924, Senator Arthur Capper of Kansas spoke to a congressional subcommittee concerning divorce. Capper stated that figures indicating that one of six marriages ended in divorce proved that Americans were abusing the availability of divorce. If this misuse continued, the institution of the family would gradually disintegrate. If the family disappeared, "the nation itself must suffer, since the family is the recognized fundamental unit of all civilization."[18]

On January 23, 1924, Capper introduced a constitutional amendment that would allow Congress to pass divorce legislation. Capper also introduced a divorce bill stipulating that divorces would be granted on the grounds of adultery, physical and mental cruelty, abandonment or failure to provide for one year or more, incurable insanity, or the commission of a felony. Notices of impending divorce suits would have to be served in person rather than by publication in a newspaper. And a year would have to elapse before a divorce became final and the parties could marry again.

When words of support poured in from religious organizations, bar associations, and concerned individuals, Capper must have enjoyed a brief fantasy in which he played moral hero to the nation. But other people soon voiced opposition to his plan. State leaders continued to resent any hint of federal intrusion into their right to regulate the marital status of their citizens. Many feminists feared that uniform law would take away the protection divorce gave women and that in the "incompetent" hands of male "statesmen, ecclesiastics, lawyers, and judges" uniform divorce law would become restrictive divorce law. Some feminists were appalled that leaders of the General Federation of Women's Clubs embraced such a "mischievous movement" as uniform law, and they recommended the adoption of free divorce instead. Yet other opponents of Capper's plan agreed with the governor of Maryland, Albert C. Ritchie, that harmonious divorce statutes were impossible in "a country of one hundred and ten million people, who include fourteen million of foreign birth" and a vast number of African Americans, all "residing throughout a territory three thousand miles from sea to sea."[19]

As in previous years, the uniform divorce law proposal failed to gain enough support in Congress to come to a vote. Still, advocates of uniform divorce law continued to push their cause unsuccessfully until 1947.

## The Reno Syndrome

Proponents of uniform divorce law kept the issue of migratory divorce in the public eye throughout the pre-World War II era. They discounted statistics indicating that only 3 to 20 percent of divorces were obtained in states other than the state of marriage and that only a few of these involved spouses who had purposely migrated to obtain a divorce. Instead, they talked in terms of an "exodus" of divorce-seekers from New York and an "interstate migration" of British Columbian divorce-seekers into Washington state.[20]

Of course, Nevada was the most highly publicized destination for divorce-seekers. The state, and especially the city of Reno, soon gained a reputation as a jurisdiction that combined lax laws, leisure pursuits, and a pleasant climate. Reno began its rise to infamy as a divorce mill largely because of Nevada's six-month residency requirement for citizenship, voting, and divorce, a provision intended to accommodate the needs of a highly mobile population of miners and entrepreneurs. In 1900, a well-known Englishman, Lord Russell, divorced his wife in Reno, married another woman, and was subsequently sued for adultery by his first wife.[21] The resulting scandal drew widespread attention to Nevada's lenient divorce laws.

Other well-known people soon took advantage of Nevada's six-month residency requirement and permissive grounds for divorce, including a broad, catch-all cruelty provision. In 1905, the ease of Reno divorce, at least for those who could afford to travel to Nevada and spend six months there, was brought to public attention by Laura B. Corey of Pittsburgh. Corey claimed that her wealthy husband was involved with a dancer. The resulting publicity catapulted Reno into the national spotlight.[22]

In 1907, William H. Schnitzer, a New York City lawyer, further publicized Reno when he established an office there and published a pamphlet describing Nevada's generous divorce provisions and Reno's attractions as a divorce mecca. Schnitzer also advertised widely in newspapers and theater programs, promising "quick and reliable action" and the "shortest residence" necessary for a divorce. Although he was reprimanded by the Nevada Supreme Court in 1911 by a temporary suspension of his license, the divorce industry had taken hold in Reno.[23]

Within a few years, the inhabitants of Reno were embroiled in a struggle that pitted the pro-divorce advocates against the anti-divorce faction. The pattern of resistance to easy divorce followed that in earlier divorce mills, including Sioux Falls, South Dakota, Fargo, North Dakota, and Guthrie, Oklahoma Territory. Women's groups, including church societies and mothers' clubs, circulated anti-divorce petitions. Ministers sermonized against easy divorce. And in 1913, Reno's major newspaper, the *Nevada State Journal,* insisted that, "The state and this city cannot advance permanently unless they be fortified not only in self-respect, but in the respect of all who think of us. Any work too damaging for any other state to do is certainly too damaging for Nevada to do."[24]

Later that week, on February 7, 1913, one hundred and sixty anti-divorce protestors boarded the Virginia and Truckee Railroad train and traveled from Reno to the Nevada state capital in Carson City. There they thronged into the state capitol demanding that a bill requiring one year's residence of all divorce petitioners be expedited through the legislature.[25]

After the one-year residency requirement passed into Nevada law, businesspeople and entrepreneurs created a public outcry against the measure. They sorely missed the revenues, including transportation costs, legal fees, entertainment, meals, and lodging, that the divorce trade put into their pockets. When the legislature convened in January 1915, the Reno Businessmen's Association requested reinstatement of the six-month residency requirement. Although women's groups, clergy, and yet another delegation from Reno protested such an action, a six-month residency bill passed on February 17, 1915. Although Governor Emmet D. Boyle believed that the issue should go to a popular referendum, he signed the bill.[26]

The matter of brief residency was far from dead, however. Opponents of easy divorce raised the question of a one-year residency bill again in 1922, but voters rejected the idea by three to one. Five years later, Nevada reduced its residency requirement from six to three months in a bill pushed through the legislature and quickly signed by Governor Fred B. Balzer before protest arose. The *Nevada State Journal* responded in a scathing editorial: "One of the most amazing legislative performances of record happened early yesterday morning when a three months divorce law was jammed through and given executive approval before the public had any inkling of the proposal. . . . Such procedure is repugnant to the principles of free government and forms a most dangerous precedent."[27]

Always under pressure from other divorce mill states, especially Idaho and Arkansas, Nevada reduced its residency requirement again in 1931. When it appeared that the other two states were about to match Nevada's three-month requirement, Nevada dropped its provision to six weeks. The *Nevada State Journal* declared that: "Revival of Gold Rush Days Predicted. Best This One, If You Can." Governor Balzer again signed the bill.[28]

The six-week residency requirement was accompanied by a number of other provisions designed to attract divorce-seekers to the state. A court could grant a divorce to a plaintiff on the forty-third day of residency if the defendant filed an appearance through an attorney. In uncontested cases, a court could accept the testimony of the plaintiff without corroborating witnesses. And liberal gambling provisions guaranteed that divorce-seekers would be able to amuse themselves— and spend more money—while they waited for their residencies to become final. The *Nevada State Journal* now seemed reconciled, but cautious. "No one denies that the 'business of divorce' has brought millions of dollars into the state annually," it admitted, but it also warned that if other states, especially New York, liberalized divorce statutes, economic disaster might follow.[29]

During the 1920s and 1930s, Reno's divorce business increased dramatically. In 1926, Nevada courts granted 1,021 divorces. After the residency requirement was lowered in 1927, this figure almost doubled to 1,953 that year. In 1928, Nevada courts granted 2,595 divorces. In 1931, when the requirement dropped to six weeks, the figure doubled again, jumping from 2,609 in 1930 to 5,260 in 1931. During the 1930s, however, Las Vegas's glitzy image and glamorous casinos drained off some of Reno's divorce trade. By 1940, Nevada's divorce rate, forty-nine divorces per one thousand people, was by far the highest of all the states.[30]

Still, we must be careful to avoid exaggerating the importance of migratory divorce. In 1940, Nevada divorces accounted for only one out of fifty divorces granted in the United States. Because famous and wealthy people sought divorces in Nevada, with all their attendant publicity, the state's easy divorces became enlarged in the public mind. Easy Nevada divorce became the symbol of the ills plaguing the American family and novels, plays, and films, including *The Women* and *The Misfits*, kept the issue in the public eye.

Three veterans of Reno divorces were among those interviewed for this study. One woman, now eighty-seven years old, obtained a Reno divorce in 1936. At the time, she was an accomplished journalist and fiction writer who could afford to spend six weeks establishing a residency in Nevada. Far from playing while in Reno, however, she continued to write. In fact, she wrote a widely circulated piece on Reno divorce while in Reno waiting for her own divorce. In her interview, she confided that the cause for her divorce was alcoholism and adultery on the part of her actor husband, but she wanted to avoid admitting to adultery in her home state of New York. In Reno, she took the easy route: she pled cruelty.[31]

A second female respondent, aged seventy-three, stated that she sought a Reno divorce because she had fallen in love with a man other than her husband. According to her, her husband had committed no marital misdemeanors. When he realized that she no longer loved him, he helped fund her six weeks in Nevada. They had just moved to Iowa and it would have taken her a year to establish residency there. She lived in a boardinghouse in Reno in 1938, then charged her husband with cruelty. Fortunately from her point of view, Nevada divorce law did not require that she offer any substantiating evidence of her charge.[32]

The third respondent was male, aged seventy-six. He revealed that his wife had engaged in an affair with his uncle, who lived in the house with them, while he was at work. He was unsure whether his son was his or his uncle's. When discovered, his wife refused to end the affair and eventually moved with this uncle to another state. In 1938, he obtained a Reno divorce on the charge of cruelty. He had ignored the matter of divorce for two years, but now wanted a quick divorce so he could remarry. His intended wife accompanied him to Reno and they married there the day after he obtained a divorce. In his case, unlike that of the previous respondent, he could have presented a catalog of substantiating evidence.[33]

These cases indicate that both famous and not-so-famous people sought divorces during Reno's heyday as a divorce mecca. As Reno became equated with easy divorce in the public mind, divorce reformers grew increasingly outraged. They were primarily upset that migratory divorce-seekers subverted the laws of their home states and territories by obtaining divorces in more lenient jurisdictions such as Reno. But, early in the twentieth century, it was becoming apparent that migratory divorces were also creating sticky legal situations when divorce-seekers from strict jurisdictions returned home. Was a Reno divorce gained in six months on one of seven loose grounds valid in New York where adultery was the sole ground for divorce, or in South Carolina where divorce was prohibited entirely?

Anxious to maintain legislative and judicial consistency in a country composed of disparate states, the nation's founders had tried to look ahead to such situations. They had written a "full faith and credit" into the United States Constitution, which went into effect in 1789. On May 26, 1790, members of the new United States Congress passed a "similar" act. According to this legislation, each individual state was obligated to recognize "the public Acts, Records, and Judicial Proceedings of every other state" in the union. If an action was valid in the jurisdiction where pronounced, it was theoretically valid everywhere in the United States; if it was questionable in its own jurisdiction, it was open to question elsewhere.

In the case of divorce actions, full faith and credit turned out to be a difficult principle to apply. Because each individual state had the right to regulate the marital status of its citizens according to its own standards of morality and public policy, each enacted its own divorce statutes. Because some states' provisions were incompatible with those of other states, one state was often unwilling to grant full faith and credit to the other. Being called upon to grant full faith and credit to divorce provisions of another state could force a state to deny its own morals and policies. If, for example, a particular state allowed divorce only on the ground of adultery, must that state give full faith and credit to a decree obtained by one of its citizens in another state offering the more lenient ground of mental cruelty? If a state had to accept the more permissive cruelty decree under this doctrine, was it not being forced to expand its own limited grounds against its will and original intent?

Yet chaos might ensue if credit was withheld from divorce decrees obtained in other states. If a citizen of a strict state obtained a divorce in a lenient state and then returned to the home state to remarry, he or she might be committing bigamy. If the home state refused to give full faith to the divorce obtained in another state, the person's first marriage would be considered valid and the second marriage bigamous. This situation would make children of the second marriage illegitimate. And it could lead to both a first and second wife claiming the property of a twice-married husband.

Migratory divorce cases occasionally caused a head-on collision between the full faith and credit doctrine and the principle that a state could regulate the marital status of its own citizens. From time to time, a spouse whose mate obtained a migratory divorce challenged the validity of that decree in the courts of his or her home state. In other cases, one spouse filed a divorce petition in the couple's home state at the same time that the other spouse applied for a decree in another state. This raised the question of which action should predominate.

Decisions in such cases were inconsistent. As early as 1813, Stephen Fitch of Connecticut obtained a Vermont divorce and married Rebecca Borden of New York. Borden's mother, who considered Fitch's first marriage valid, won $5000 in damages when she sued him for debauching her daughter. When Fitch appealed the verdict, the New York Supreme Court upheld the judgment, ruling that the Vermont divorce was invalid. But, in 1832, the Maine Supreme Court *upheld* a divorce obtained by a wife in Rhode Island. And in 1874, the Massachusetts Supreme Court ruled that one husband's Indiana divorce was valid, even though his wife was uninformed of the divorce proceedings.[34]

During the early 1900s, a number of similar cases came before the United States Supreme Court. In 1906, the Court established a principle that would be widely applied to migratory divorce appeals during the next thirty-six years. In the case of *Haddock v. Haddock*, the court ruled five to four in favor of Harriet Haddock's contention that her husband's Connecticut divorce was invalid. Because the couple's marital domicile was New York, the court decreed that New York courts had the right to reject an out-of-state decree.[35]

After 1906, some judges followed the principle of marital domicile established in *Haddock v. Haddock*, but others used residency requirements to guide their decisions regarding migratory divorces. Some judges felt that if basic residency requirements had been met in the venue where the divorce was granted, the divorce deserved full faith and credit. If, however, the plaintiff established residency only to obtain a divorce, the out-of-state decree should be denied full faith and credit.

As Americans became both mobile and affluent enough to escape the rigid divorce laws of their own states by temporarily locating in more permissive jurisdictions, the number of migratory divorces seemed to increase. Although statisticians continued to assure Americans that numbers of migratory divorces were reasonable because in actuality many couples married in one state and later lived in another where they might divorce, the blatant publicity garnered by Reno and other divorce meccas convinced Americans otherwise. Also, courts in strict divorce states were increasingly called upon to resolve the thorny issues that resulted from migratory divorces.

One especially complicated case reached the United States Supreme Court during the early 1940s. The situation that resulted in the case of *Williams et al. v. North Carolina* began when Lillie Hendrix and Otis Williams left North Carolina to relocate in Las Vegas for six weeks. After fulfilling Nevada's sixty-day residency requirement, they obtained divorces from their North Carolina spouses. Hendrix and Williams immediately wed each other and returned to North Carolina as a married couple. Following the earlier ruling in *Haddock v. Haddock,* North Carolina authorities refused to recognize their out-of-state decrees. They also charged the couple with bigamy and sentenced them to two years' imprisonment. After the Supreme Court of North Carolina upheld the bigamy conviction, the couple appealed the judgment to the United States Supreme Court. In 1942, the Supreme Court ruled that because due process had been followed, the Nevada decrees deserved full faith and credit. The Supreme Court thus overturned the principle of *Haddock v. Haddock.* Justices Robert Jackson and Frank Murphy strongly objected to this decision on the ground that it repealed "the divorce laws of all the states and substitutes the law of Nevada" in cases in which spouses could afford to temporarily relocate in Nevada.[36]

Many judges subsequently attempted to follow the precedent established in this 1942 United States Supreme Court decision. In 1943, the case of *Lambert v. Lambert* challenged the validity of a Nevada divorce decree in the state of New York. The Lamberts had married in New York in 1920 and separated twenty-one years later. John Lambert then relocated in Nevada, petitioned for a divorce on the ground of mental cruelty, married another woman on the day the divorce was granted, and moved to Massachusetts. But Lambert's wife, Beatrice, had filed a petition for divorce on the ground of adultery before Lambert left for Nevada. Following the 1942 *Williams v. North Carolina* decision, the Monroe County court in New York ruled that the Nevada divorce deserved full faith and credit because Lambert had met Nevada's legal requirements for a divorce. The court thus blocked Beatrice Lambert's attempt to obtain a New York divorce—and a more favorable property settlement.[37]

In the meantime, the state of North Carolina persisted in prosecuting Hendrix and Williams on a slightly different issue. The state argued that the couple had failed to establish a *bona fide* residence in Nevada and thus their Nevada divorces were invalid. In 1945, the Supreme Court examined whether six weeks' residency in the Alamo Auto Court in Las Vegas constituted valid residence. This hearing explored the question of whether North Carolina courts had to give full faith and credit to a divorce when the petitioners had failed to establish a *bona fide* domicile in Nevada. In 1945, the Supreme Court ruled that because residency in this case was a sham, North Carolina was free to deny full faith and credit to the Nevada decrees.[38]

The second *Williams* decision established the principle that if a divorce-seeker established residence in a state only to acquire a divorce, the resulting decree was not entitled to full faith and credit by other states. If, however, a petitioner established a genuine domicile, the divorce decree was entitled to full faith and credit in other jurisdictions. Although this ruling came too late to help Beatrice Lambert, it did affect other litigants. In 1946, the case of *Crouch v. Crouch* came before the Supreme Court of California. Edith M. Crouch challenged the validity of Ben E. Crouch's Nevada decree by entering her own divorce petition, which included a claim on their property. The court found that Ben Crouch had established bogus residence in Nevada; he intended to become a Nevada citizen only until he secured a divorce *and* a more favorable property settlement than offered under California law. Because the California court regarded Ben as Edith's legal spouse despite his Nevada decree, it granted her the right to file a divorce petition, including a property claim, against him.[39]

After 1945, then, a general principle regarding migratory divorce prevailed: according to the Constitution, congressional action of 1790, and subsequent judicial rulings, a valid divorce decree granted in one state was entitled to—but might not always receive—full faith and credit in other states. Although it was generally true that a decree that was valid in the jurisdiction where it was rendered was valid everywhere, that validity was not guaranteed: a home state's laws could also be used to determine validity.[40]

The full faith and credit doctrine also generally excluded divorces granted by jurisdictions outside the United States. No state was required to recognize the validity of a divorce granted by another country, although most frequently did so on the principle of comity— courtesy to another jurisdiction's laws. But a state had the right to reject a decree that offended its moral standards or opposed its public policy. As other destinations became attractive to highly mobile Americans, it was increasingly unsafe to assume that a quick divorce granted in Mexico, for example, would hold in the courts of the petitioner's home state or territory.[41]

Clearly, the right of individual states to regulate the marital status of their citizens restricted the application of the full faith and credit principle in cases of migratory divorce. As Justice Murphy argued in his dissenting opinion in the first *Williams* appeal, divorce actions involved the interaction of constitutional principles and state policy.

Murphy argued that because "marriage and the family have generally been regarded as basic components of our national life," the solution of "problems engendered by the marital relation, the formulation of standards of public morality in connection therewith, and the supervision of domestic (in the sense of the family) affairs" have been left to the individual states, for each has "the deepest concern for its citizens" in such matters. He concluded that when a conflict arose between two states on matters of domestic policy, a court was unable to simply apply the constitutional principle of full faith and credit; it had to take moral values and public policy into account as well.[42] This interpretation has guided the application of full faith and credit in divorce actions until the present day.

Unfortunately, jurists failed to develop a similar guiding principle for the growing problem of collusion. Essentially mutual agreement between spouses to seek a divorce, collusion had long been against the law in most states. Judges could even deny a divorce if collusion existed. Still, legal experts agreed that many divorcing couples practiced it and judges tended to overlook it.

Collusion subverted the adversarial nature of divorce actions in which an innocent spouse sued a guilty spouse. Instead, couples who agreed between themselves to "sue" for divorce, usually also decided whether it would be more convenient and seemly if the wife or the husband was the plaintiff. They might also take into account the costs and amount of court time involved if one or the other initiated the suit. Because women received more divorces than men, it is reasonable to assume that in cases of collusion, couples agreed that wives should be the plaintiff more often than husbands. During the 1930s, a Richmond, Virginia, man who was considering divorce said as much: "Custom demanded, to a large extent, that the husband allow the wife to make the first move."[43]

The contention that many couples agreed which party would be the plaintiff is borne out by a sample of forty-five midwestern women and men who obtained divorces before 1945. Thirty-seven interviewees revealed that they and their spouses had agreed that the wife should be the plaintiff. They believed that if she accused him of fault instead of him accusing her, she would bear less stigma in the eyes of their children whom she would be raising. They also thought that if a wife appeared as the wronged party, this would encourage a court to accept her property and child custody demands—arrangements that the couple had already agreed upon. Eight other couples chose the husband as plaintiff, but none of these cases involved children so neither stigma nor custody settlements was an issue. Five of these couples decided that the husband should act as plaintiff because the wives feared dealing with attorneys and appearing in court. One husband was willing to take care of all legal matters and hire an attorney friend at a special rate. Only years later did his former wife accidentally learn that he had moved out of town without paying his "friend."[44]

Certainly, many migratory divorces had a collusive element. When spouses agreed that one of them would go to a lenient jurisdiction to obtain a divorce, they were practicing collusion. And when a spouse promised to refrain from bringing a counteraction or attempting to revise the terms of a migratory divorce, she or he was being collusive.

The state of New York provided the setting for another type of collusion. Because New York specified only adultery as a ground for divorce, many people created sham adultery cases. Companies sprang up that supplied a hotel room, a phony partner, a private detective, and a photographer. Once the incriminating photographs were taken, the case went to court where the partner and detective testified to a husband's or wife's adultery. During the early 1920s, one entrepreneur used unemployed actors in producing fabricated court evidence.[45]

In other states, collusion frequently went hand-in-hand with the use of moderate grounds. Few wives and husbands who agreed to divorce wanted to stain their spouses' reputations with charges of adultery, alcoholism, or impotency. Even if accurate, such harsh grounds could easily be replaced by the less damaging grounds of cruelty, neglect of duty, or incompatibility when they were available. Of the forty-five collusive couples mentioned above, forty-two chose such grounds as cruelty, neglect of duty, or incompatibility, although in at least twenty-nine of these cases, adultery and alcoholism had destroyed their marriages.[46]

In flaunting the adversary system, collusive couples practiced a form of what was later called no-fault divorce. Evidently, the time was not yet ripe for the acceptance of no-fault divorce; few people even raised the possibility. Instead, collusion remained in force, thus establishing in practice what later no-fault legislation would recognize by statute.

### The Search for Causal Factors of Divorce

As the divorce rate rose and problems concerning full faith and credit as well as collusion captured the public's attention, a growing number of Americans felt compelled to identify factors causing divorce. If they understood causes of divorce, they reasoned, they might be able to curb divorce, or at least deal effectively with associated problems. And if they were aware of causal factors, policy-makers and legislators could devise rational solutions for such problems as the rising divorce rate, migratory divorce, and collusion.

The move to detect causes of divorce was hardly new, but it intensified as the divorce rate climbed. Commentators ranging from sociologists to novelists to judges posited their pet theories regarding divorce.
Like earlier commentators, several writers linked American individualism and democratic ideals to the escalating divorce rate. In 1915,
one advocate of easy divorce declared that the growth of divorce in
the United States signaled Americans' "increasing individualism" and
their "demand for a larger degree of freedom and happiness."[47]

In 1921, a best-selling novel elaborated upon the tie between
divorce and individualism. In a plot full of twists and turns, Charles
Norris's *Brass: A Novel of Marriage* described Philip's shift from an anti-
divorce to a pro-divorce position. When Philip first discovered the
darker side of marriage, he found divorce repugnant: "People he
considered decent did not get divorces." Then Phillip's wife divorced
him. Philip eventually remarried, but was again unhappy. When a
minister told Philip that "civilized society" could not endure the existence of marriage bonds that were "dissoluble at will in the divorce
court," Philip balked; he had come to believe that a democratic country should freely allow divorce. Norris, who was evidently thinking of
the Declaration of Independence, had Philip proclaim that the United
States "Constitution specifies 'liberty and the pursuit of happiness' for
all."[48] Although confused about the document that stated the principle, Norris believed that a dedication to liberty and the spread of
divorce went hand-in-hand.

Other commentators argued that the rising divorce rate was
caused, at least in part, by people's rising expectations of marriage. In
1929, for example, Robert and Helen Lynd broached this idea in their
famous Middletown study. The Lynds' view of divorce has recently
received a good deal of development and testing. A study comparing
divorce petitions filed in Los Angeles during the 1880s and in 1920,
for example, has explored petitioners' expectations concerning companionate marriage: a union that offered friendship, love, romance,
happiness, sex, and personal fulfillment for both spouses. Still other
studies have disclosed that Americans' expectations of courtship,
women's roles and work, and family life underwent significant modifications by 1945, all accompanied by significant demographic changes
as well.[49]

During the early twentieth century, other experts, both genuine
and self-styled, theorized that a wide variety of other factors caused
the rising divorce rate: increased median income, industrialization,
decline in economic functions of the family, weakening of religious
tenets, and fading social stigma concerning divorce. In particular,
changes in women's roles were blamed for divorce. Instead of staying
at home full-time, many women held paid employment outside their

home. They also demanded participation, especially the right of suffrage, in the political realm. Emancipation supposedly gave women "ideas" as well as the financial resources to act upon them.[50]

Other analysts thought that it was more important than ever to understand the relationship of divorce statutes to the divorce rate. One study examined the divorce rate in fourteen states that revised their divorce statutes between 1922 and 1928. Its author concluded that the result of legislation on divorce rates was "negligible" in these states. He added that if the "well-intentioned but uninformed people" who were "grasping blindly for a remedy to preserve the American home" got their way, they would soon discover that uniform divorce law was ineffective in stemming the divorce rate.[51]

In 1931, another investigator took a different tack. Isabel Drummond compiled the divorce statutes of every state and territory in the United States. She believed American divorce statutes were too harsh; legislators were "almost despotic" in their "high-handed manner of dictating terms to disgruntled mates." Unlike such "forward-looking countries" as "Sweden, Japan, and Russia," which recognized divorce by mutual consent of spouses, the United States imposed a difficult and costly adversary procedure on divorce-seekers. Still, Drummond admitted, despite the difficulty of American divorce provisions, the divorce rate continued to rise. Perhaps, she concluded, divorce statutes—whether strict or lenient—failed to affect the divorce rate one way or the other after all.[52]

Statisticians tried to test these varied hypotheses, but divorce figures were fickle. Arranged one way, statistics proved that a particular cause spurred divorce; arranged another, they proved the validity of a different cause. In 1932, one statistician argued that no one factor caused divorce. Instead, multiple factors interacted with each other. He concluded that American "home life is so diversified, and the changes in society of six decades so complex that a singular interpretation of increasing divorce is scarcely plausible."[53]

The erratic collection of divorce statistics further hampered analysis. Despite the good intentions of federal agencies and Congress, divorce statistics were collected only intermittently after the end of the second government survey in 1906–07. Although government officials hoped to collect marriage and divorce statistics every ten years, the nation entered World War I in 1917, the year of the next scheduled survey. Consequently, the Census Bureau gathered figures only for 1917.

Between 1922 and 1932, the Census Bureau was able to collect marriage and divorce statistics each year and publish an annual report, *Marriage and Divorce*. Because of the Depression, which began in 1929, the Census Bureau did not gather statistics during the mid-

1930s. During 1939 and 1940, statistics were collected from twelve states that Census Bureau officials designated as a divorce collection area. The United States entered World War II in 1941, and data collection was again abandoned between 1942 and 1945.[54]

In addition to statistical data, divorce records themselves offer analysts another source of information. Divorce records are, however, extremely difficult to interpret. They often fail to disclose real causes of marital breakdown, and only partially reveal the realities of the people and the marriages they represent. For a number of reasons, the true and complete experiences, emotions, and motives of divorce-seekers often remain more clouded than exposed by divorce records.

Divorce documents from Linn County, Iowa, between 1928 and 1944 demonstrate the difficulties involved in analyzing divorce records as well as some of the useful and provocative information they contain. The records are plentiful, for in spite of Iowa's bucolic image, created in part by Grant Wood's paintings and the novels of Bess Streeter Aldrich and Hamlin Garland, many Iowans divorced. In 1930, the state's divorce rate was exceeded by only twenty-two other states. In Linn County between 1928 and 1944, which was a single period in county records, 4,758 divorces were granted. Of these, 3,796, or nearly 80 percent, went to women, while 962, or slightly more than 20 percent, went to men.[55]

Basic details of each Linn County divorce were recorded in a register, *Record of Divorce, 1928–1944*, that includes names, nationality, race, number of prior marriages, date of marriage, grounds for divorce, date of divorce, and whether alimony was awarded. From an analyst's viewpoint, it is unfortunate that the divorce register omits numbers of children involved and decisions concerning custodial parents. In addition, child support awards are recorded only in margin notations. And, the register neglects to specify which cases were annulments rather than divorces. Annulments can be identified only by the ground listed—consanguinity, impotence, bigamy, insanity, or non-age of either party.[56]

Further information concerning each couple listed in the register can be found in a case file, which includes a divorce petition, such court orders as writs of attachment and injunctions, a transcript of a couple's divorce hearing, and a divorce decree. These case files, which are not easily accessible to researchers, have a number of deficiencies. In particular, case file documents fail to routinely record the presence or absence of children and child support awards. One wife received custody of the couple's infant daughter, but no mention is made of child support although she had specifically requested it. In another case, the register indicated that child support had been awarded, but case file documents give no indication that a support order existed.[57]

Another problem with Linn County case files is their failure to record the occupation of litigants except in an occasional passing mention. In asking for custody of a seven-month-old daughter and support for the child, one wife stated that her husband worked at a starch factory in Cedar Rapids where he earned $5 a day. Another wife testified that her husband earned $165 per month working as a railroad switchman, and one that her husband worked for the Lock Joint Pipe Company in Kansas City. References to women's employment were even more sparse, but occasionally a husband remarked that his wife worked as a domestic or in another job. In other cases, wives noted that they had worked during the marriage, but seldom specified their jobs.[58]

In addition to discrepancies and omissions, case files frequently lack enough information for a reader to draw a reasonable conclusion about certain cases. In one sketchy set of records, a Bohemian woman charged her husband with cruelty, but gave no details or substantiating incidents from their six-year marriage.[59] As a result, the occurrences that drove her toward the divorce court are unknown.

In a similar case, a husband petitioned for divorce from his wife on the ground of cruelty, but failed to offer substantiating examples. Yet, on April 10, 1943, only nineteen days before their twenty-seventh anniversary, he received a divorce. Although his wife was the guilty party, the court awarded her $10 alimony per week and title to their house and household goods. The records do not indicate whether the husband offered this settlement, if the wife asked for it, or if the judge initiated it.[60]

Case file documents also leave a good deal unexplained in another far more intricate suit that eventually resulted in two divorce decrees. The husband, who had been married four previous times, and the wife, who had been married six previous time, had obtained a divorce on January 16, 1940, on an unspecified ground. They remarried on February 23, 1940. In late 1940, the wife applied for a divorce, testifying that her husband beat her, cursed her, and called her "all manner of vile names." Her husband responded that she drank to excess, had threatened to kill him, had thrown dishes at him, and had "made a general nuisance of herself." The judge awarded the wife a divorce and one $50 alimony payment. A few days later, on August 30, 1940, the former wife and husband were back in court asking to have the decree set aside. The husband then petitioned for a divorce from his wife. On December 4, 1940, the same judge ruled for the husband.[61] One can only guess whether this was a case of a tempestuous relationship with numerous break-ups and reconciliations, or one in which a vindictive husband tricked his wife in order to save face and $50 by getting a divorce himself.

Clearly, it is difficult to draw accurate conclusions about divorce petitions and decrees from such inconsistent, incomplete data. Consequently, answers to important questions concerning child support and which occupational groups most frequently utilized divorce remain uncertain. In addition, it is almost impossible to divine people's true reasons for divorcing their spouses. Linn County divorce-seekers had a choice of only six grounds during this period. They chose the ground that most closely approximated their situations, was the least damaging to everyone concerned, or was the most likely to persuade a judge to give them a divorce and a favorable settlement. Consequently, their divorce petitions were legal documents rather than honest, complete stories about their marriages.

In 1928, Iowa's six well-established grounds for divorce were: adultery, desertion for period of two years, conviction for a felony after marriage, habitual drunkenness after marriage, such cruel and inhuman treatment as to endanger life, and prior pregnancy of wife by a man other than the husband.[62] Because virtually no Linn Countians used the charge of prior pregnancy, they essentially had only five grounds available to them. Whatever conditions terminated their marriages, petitioners had to fit them into the fabric of these charges.

Linn County's divorce register book between 1928 and 1944 indicates that petitioners' choices of grounds were very similar to the choices made by Americans nationwide. The majority of Linn Countians leaned toward broad, relatively non-damning grounds, especially cruelty. It is likely that a number of these divorcing couples colluded in choosing this ground rather than harsher grounds, including adultery and alcoholism. Although collusion concerning grounds can seldom be proven, the possibility of its existence demands that records be read with a critical eye.

During this period, the majority of Linn Countians cited the charge of cruel and inhuman treatment in their divorce petitions. In a randomly selected sample of 900 cases from the divorce register, 85 percent of petitioners pleaded cruelty; nearly 88 percent of women used this charge, while nearly 78 percent of men did so.[63]

Because the "cruel and inhuman treatment" clause included the restriction that a spouse's health must be impaired by such treatment, most petitioners focused upon life-threatening aspects of a spouse's behavior. Although several court decisions had ruled that the cruelty charge could include verbal abuse and such psychological mistreatment as making unfounded accusations of unchastity against a mate, most petitioners seemed hesitant to take a chance on the mental cruelty plea. Instead, they regularly described slapping, punching, beating, and stabbing. Even those who charged verbal cruelty usually bolstered it by recounting a slap or two.[64]

One of the few petitioners who took advantage of the enlarged construction of cruelty was a German man who sued his German wife for divorce in 1928. He charged cruelty because she had falsely assured him that she had sufficient funds to convert the top floor of his house into a rental unit. He married her, engaged plumbers and other workpeople, and presented the bills to her. She stalled by saying that she preferred to withdraw her funds after the end of the current interest period. Eventually, she admitted that she had lied about having capital to invest in the house. She then refused to co-sign a bank loan to pay the bills. A subsequent divorce decree ensured that his house would be free from any claims by her.[65]

The second most popular ground was desertion. Of the sample of 900 cases, 10 percent of Linn County petitioners claimed desertion; 8 percent of women and nearly 19 percent of men did so. In one desertion case, an African American woman sued her African American husband for divorce in 1930, seven years after he deserted her and their two children. Although the documents fail to disclose her reasons for waiting so long to obtain a divorce and the judge's reasons for ruling that she could immediately remarry, it seems probable that she had recently become romantically involved. Marriage would give her children a home, a high priority for most judges who often allowed immediate remarriage in such cases.[66]

An especially curious case of desertion came before the county in 1927 when a mother of a ten-year-old boy accused her husband of refusing to supply her with "sufficient necessaries and comforts," being habitually drunk, and, by his behavior, forcing her to leave him. Because he had never asked her to return, she argued that "such conduct on his part amounted to legal desertion." The court apparently agreed, for it granted the divorce.[67]

Only slightly more than one percent of the sampled Linn County petitioners used the ground of conviction for a felony; 1.5 percent of women and 0.5 percent of men used this charge. Felony cases tended to be cut-and-dried, probably because the evidence of a spouse's incarceration was indisputable. One wife with three young children under the age of three testified that her husband was in an Iowa reformatory because he had committed an unnamed felony. She requested custody of the three children, $50 per month child support, and permission to remarry before the end of the one-year waiting period. She got custody of the children, but her husband's argument that he was unable to earn money while in jail was a telling one; she received neither alimony nor child support. The divorce decree also omitted any mention of remarriage.[68]

Another one percent of sampled Linn County petitioners accused their mates of being habitual drunkards; nearly 2 percent of women and 0.5 percent of men used this ground. In one instance, a mother of three young children was divorcing her husband for the second time. She explained that she had remarried him because he had promised to quit drinking. But he continued to drink heavily; he went on "sprees" and beat her. She received a divorce and custody of the children, but no child support. In another case, a husband who stated that his wife drank, cursed at their three daughters, and struck him received a divorce and custody of the children.[69]

Slightly less than one percent of sampled Linn County petitioners charged adultery; .56 percent of women and 1.6 percent of men did so. In a 1930 case, for example, a wife charged her husband with abandoning her and their two children, threatening her with violence, refusing to support her and the children, and committing adultery with a woman whom he illegally "transported" from Illinois to California for "immoral purposes." She won a divorce, custody of the children, and $10 per week child support.[70]

Husbands also accused wives of adultery. In one especially harsh petition, a husband charged his wife with adultery and with having been "an inmate of a house of prostitution" during recent months. He received custody of their three children, but the judge allowed her visitation rights. Although it may seem curious that the judge allowed a woman of her reported character to visit her children, many judges believed that even errant mothers had important bonds with their offspring.[71]

In the Linn County sample, every petitioner cited specific charges of cruelty, abuse, and desertion. Although such factors as rising expectations of marriage, a decline in the patriarchal family, or women's expanded employment opportunities may have contributed to the disintegration of their marriages, petitioners had to frame their charges to fall within the limits of Iowa divorce law rather than exploring and recording their full range of reasons for requesting a divorce.

In addition, it must be remembered that divorce-seekers typically presented only one side of the story. In most cases, little opportunity occurred for defendants to present their own complaints unless they counter-sued. Also, petitioners were often likely to paint the worst possible picture of their marriages and spouses in order to persuade a judge to grant them a divorce and award them a favorable settlement.

Conflicting testimony in contested divorce suits demonstrates the folly of taking a petitioner's view of marital failure and his or her spouse's sins as truth. In 1928, a woman who had been married six times before sued her husband who had been married twice before on the ground of cruelty. She testified that he cursed at her and refused to give her money. In his cross-petition, her husband maintained that she nagged, complained, and wanted a home in Cedar Rapids, rather than in Clinton, where he worked. He added that she cursed at him and harassed him to the point that it impaired his health. She got a divorce and $200, while he paid court costs and attorney fees.[72]

Another 1928 case that demonstrates the principle of two sides to every story involved a husband who stated that because his wife associated with other men, much to his "humiliation and grief," he wanted a divorce, their house, automobile, and all household goods. His wife responded that because she had been employed and had done all the housework as well, she was entitled to an equal portion of their property and household goods. She added that her husband had physically abused her, made her so nervous that it "spoiled her peace of mind and her appetite," cancelled her charge account, and caused her to lose her job. On May 22, 1929, the husband received a divorce and their property, automobile, and household goods.[73]

In other contested cases, defendants protested charges so effectively that they, instead of the original plaintiffs, received divorces, and favorable settlements from courts. In 1929, an African-American woman sued her African-American husband with striking, abusing, and choking her. She asked for alimony and a writ of attachment against her husband's property. Her husband denied her allegations and charged her with co-habiting with other men and currently living with an Iowa City man. After a protracted hearing, the judge ruled for the husband. In another case, a husband who had been married twice before charged his wife, who had been married once before, with calling him a "son of a bitch" and a "dirty dog," neglecting her household duties, and threatening to strike him with a poker. She countered that he had gone out with other women, knocked her into a ditch, and attacked her at a dance. The judge ruled for the wife.[74]

Despite their many discrepancies and difficulties, however, divorce records frequently offer useful, credible information to analysts. The Linn County register, for example, affirms the growing democratization of divorce, a process that was occurring all over the nation.

According to the register, people from many ethnic and racial groups sought divorces. Although the majority of Linn County litigants listed their nationality as American, others apparently failed to understand the concept because they responded that their nationality was Yankee, Iowan, Kansan, Missourian, Jewish, white, "redskin," or unknown. Bohemians, who were a sizable group in Cedar Rapids, were the dominant ethnic group to request divorces in Linn County; they accounted for approximately 5.5 percent of sampled litigants. Germans composed slightly over 5 percent of sampled litigants, while English and Irish numbered nearly 4 percent each. Handfuls of other litigants identified themselves as African, Afro-American, Assyrian,

Belgian, Canadian, Czechoslovakian, Danish, Dutch, French, Greek, Hungarian, Italian, Lithuanian, Mexican, Norwegian, Polish, Russian, Scotch, Serbian, Scandinavian, Scotch-Irish, Swedish, Swiss, Syrian, and Welch.[75]

Race was not always clearly recorded in the register, but it appears that sixty-one African-American couples divorced between 1928 and 1944. Wives obtained forty-three of these divorces, husbands eighteen. Because Iowa, unlike many other states, allowed inter-racial marriage, two inter-racial couples obtained divorces. In both cases, white wives sued black husbands.[76]

Another significant finding is the number of prior marriages of each litigant, which hints at the emergence of multiple divorce in the United States. Among female litigants in the sample, slightly less than 18 percent had been married once before, 3.6 had been married twice before, and .86 had been married three times or more before. Among male litigants in the sample, approximately 18 percent had been married once before, 4 percent twice before, and .86 three or more times before. One woman and one man admitted to six prior marriages.[77]

It is difficult to determine whether death or divorce terminated these litigants' previous marriages, but surely some ended in divorce. In 1938, an article in *Harper's Monthly* confirmed that multiple divorce existed. Its female author, who had recently married for a third time, analyzed her two divorces and speculated on reasons for multiple divorce.[78] Although multiple divorces would have skewed the divorce rate, analysts generally overlooked the phenomenon. The history and development of multiple divorce is still largely unstudied.

The Linn County case sample also indicates that a substantial number of marriages was extremely brief in duration. Twelve percent of sampled divorce cases involved marriages of less than two years. One woman requested a divorce from her husband on the ground of cruelty because he had, according to her, used violent language, threatened her life, and impaired her health by constantly pointing out her faults. They had married on October 25, 1932, and were divorced on November 8, 1933. And in 1939, a Caucasian wife charged her American Indian husband with cruelty because he had "damned and vilified" and struck her. They had married on January 12, 1938, and were divorced on September 30, 1939.[79]

Yet other data from the divorce register demonstrate that, as in most regions of the United States, alimony and child custody awards lay in the hands of judges and were something of a lottery. Usually, women received such awards far more often when they were plaintiffs—in other words, the innocent party in the suit.

In the Linn County sample, only 29 percent of litigants received alimony, 4 percent received child support in place of alimony, and 67 percent received neither alimony nor child support. In cases involving alimony awards, 87 percent of plaintiffs were women.[80]

In addition to data from the divorce register, individual case files are also useful. They flesh out information found in the register and expose aspects of divorce overlooked by it. For instance, case files show that a number of Linn Countians remarried in less than one year, the waiting period required by Iowa law. According to the documents in seventy-seven randomly selected case files, fourteen people expressed interest in circumventing the one-year requirement. In one instance, a wife and mother of two children accused her husband of "consorting" with another woman and refusing to support his family. One June 20, 1944, she received a divorce, custody of the children, and $6 a week child support. Six days later, she was back in court, claiming that her former husband had "absconded from the state" and was not making his support payments. She asked the court's permission to marry an employed Cedar Rapids man who would provide a home for her and her children. There is no evidence that the judge raised any questions about the timing of the incidents in this case.[81] It is likely that he was pleased to see a new family in the making and thus gave her permission to remarry immediately.

A number of other cases also suggests that many judges were more concerned about creating new families and providing two parents for children than they were about the legalities of the prescribed waiting period. A 1932 decree that gave a father custody of the couple's two children also gave him permission to remarry at any time, although there is no evidence in the records that he requested such a waiver.[82]

In a 1942 divorce suit, the court also condoned rapid remarriage. In this case, a wife sued her common-law husband for divorce on the ground of alcoholism. For an unexplained reason, the judge ordered that the couple's two children would remain in the custody of the juvenile court until further notice. The divorce was granted on October 5, 1942. Although neither party received permission to remarry, the wife reappeared in court on March 4, 1943 to inform the judge that she had married a defense plant worker from Burlington, Iowa. She also requested custody of the two children. Again without questioning the timing of these events, the judge remanded custody to her and her new husband.[83]

Unarguably, the Linn County divorce register and case files for the years 1928 to 1944 provide useful data and intriguing stories regarding divorce during these years. Still, analysts must use them with caution because they only partially reveal the underside of marriage. To take them at face value is to ignore their difficulties—and to distort conclusions about the factors that led to divorce.

By the mid-1940s, then, divorce was a permanent feature of American life. It *was* everywhere; it had even become a staple theme of popular literature in the United States. Amelie Rives, who had herself divorced during the 1890s, portrayed lovely heroines who graced Virginia plantation houses, while divorce closed in upon them. And Mari Sandoz explored the issues of wife abuse and divorce in the West, especially in her autobiographical novel *Old Jules.*[84]

Between 1910 and the mid-1940s, a few divorce-related issues had been resolved. The Supreme Court had hammered out a principle regarding migratory divorce. And the uniform divorce law movement had neared its end. But many problems continued to exist. The growth of collusion was subverting the word and intention of divorce statutes. Causes of divorces were still difficult to determine. And such issues as alimony and child custody continued to receive little public airing. Clearly, post-1945 America would have no lack of causes to debate and pursue.

And what of Sara Bard Field and Charles Erskine Scott Wood and their brave experiment in living together without marrying? Sara recalled that when Wood's wife died in 1933, she and Charles did not even consider the idea of marrying because they believed that "neither church nor state should interfere in a relationship" unless it involved other people. Because they had no children of their own, they saw no reason "to seek legal sanction" for their long-established relationship.

In 1937, when he was eighty-five, Charles had a serious heart attack. His brush with death made him realize that Sara might be confronted with difficult problems regarding insurance and inheritance after his death. Consequently, they renounced their vow to avoid marriage and wed on January 20, 1938.[85]

Like many other Americans, Field and Wood had both put marriages asunder and set themselves, their spouses, and their children adrift in a sea of uncertainty. They had done so for love of each other rather than marital misdeeds on the part of their mates. In a democratic nation, such decisions were possible. Although Wood died in 1944, Field lived until 1972, long enough to realize that the options that she and Charles had selected during the early 1900s would become popular as well as problematic ones during the latter part of the century.

# *The Revolving Door*

☐ When World World II ended in 1945, divorce was becoming increasingly common in the United States. With spouses frequently coming and going, the institution of marriage was beginning to somewhat resemble a revolving door. But divorce was not only ubiquitous; it was widely accepted as a traditional—or customary—way of resolving marital disharmony.

As the century progressed, the spread of divorce affected American law and society on many levels. By 1970, every state in the union permitted divorce. In that year, California adopted no-fault divorce. By the early 1980s, one out of two marriages ended in divorce. By the end of the decade, Americans divorced at the rate of over one million divorces a year—one every 13 seconds. In 1990, "Mister Rogers" commented on the impact of these changes on his popular children's television program: "If someone told me 20 years ago that I was going to produce a whole week on divorce, I never would have believed them."[1]

In fewer than fifty years, divorce in the United States experienced a revolution in its status and acceptance. Yet, when World War II ended in 1945, South Carolina and New York were still clinging to their long-standing, restrictive policies concerning divorce. South Carolina law prohibited divorce, while New York law recognized only the sole ground of adultery, but before too long, even in these states, legislators bowed to the growing need to provide their citizens with a way to terminate unsatisfactory marriages.

South Carolina's no-divorce stance fell victim to changing views of divorce in 1949. Since 1878, when South Carolina's ten-year experiment in permitting divorce ended, a growing number of cases of marital upheaval came before its courts. Requests for annulments and

156

divorces of bed and board mounted, and a growing number of South Carolinians sought resolution of alimony, dower, and property disputes after obtaining migratory divorces in other states.[2]

In 1949, legislators framed the South Carolina Divorce Act, which allowed divorce and placed jurisdiction in the Court of Common Pleas. It also specified adultery, desertion for one year, physical cruelty, and habitual drunkenness as grounds for divorce. In addition, the act established a one-year residency requirement, allowed notice of a divorce suit either in person or by publication, and required that all couples consider reconciliation. It also placed alimony and child custody decisions in judges' hands.[3]

Seventeen years later, New York divorce law underwent modification as well. The 1966 revision constituted the first alteration since 1787, the year when Alexander Hamilton sponsored a statute designating adultery as the sole ground for divorce. By the 1960s, many New Yorkers urged modification of the statute because they believed that its strictness encouraged New Yorkers to fabricate divorce cases, or seek migratory divorces. In 1966, the report of the Joint Legislative Committee on Matrimonial and Family Laws suggested a number of recommendations that legislators adopted. New York's revised divorce law specified adultery, cruelty, abandonment, confinement in prison, and living apart for two years as grounds for divorce. It also allowed both parties to remarry and estabished a tax-supported Conciliation Bureau.[4]

After the New York Divorce Reform Law went into effect on September 1, 1968, numbers of divorces rose dramatically. In 1966–67, some 4,073 divorces had been reported in the state, but in 1968–69, there were 18,180. Experts assured the public that the state's ballooning divorce rate was offset by a "reduction in perjury, fraud, insult, and disrespect for the administration of justice." Presumably, the divorce rate rose because New York couples stayed in the state to divorce rather than traveling to Reno or other permissive jurisdictions. In addition, during these years the number of annulments dropped from 2,630 to 2,123 and the number of separations from 1,982 to 802, evidence that New Yorkers preferred to divorce rather than to invent grounds for annulment, or to separate without the right of remarriage.[5]

As divorce increased, it became one of the most widely studied phenomenon in the nation. Still, although newspapers and magazines frequently reported the latest divorce statistics, the divorce rate was far from definitive. Assessing the rate of American divorce, even with the aid of computers, was a process replete with pitfalls and snares.

Despite technology and improved collection methods, divorce statistics continued to be erratic and incomplete after 1945. State variations in collection methods distorted data, as did the common practice of including annulments and divorces of bed and board in the figures. Analysts frequently complained that divorce data were "woefully inadequate" and urged the Bureau of the Census to remedy the situation. In 1958, the Census Bureau devised the concept of a Divorce Registration Area (DRA) consisting of states that adequately recorded divorce statistics, but because most states produced incomplete records, the growth of the DRA was slow. Ten years after its founding, the DRA was made up of only twenty-six states.[6]

By the 1980s, a vast array of publications were reporting divorce statistics, including Census Bureau statements and the *United Nations Demographic Yearbook*. Still, divorce statistics continued to be inconclusive because every method of analysis had its own weak point. For instance, comparing the number of divorces with the number of marriages in a given year ignored the fact that most divorces had to do with marriages contracted in previous years. When the number of marriages was high in a given year, the number of divorces in that year appeared low, and the divorce rate seemed to be declining. Conversely, when the number of marriages in the year was low, the number of divorces seemed high, and the divorce rate appeared to be rising.[7]

Another common approach to computing the divorce rate was to establish the numbers of divorces per every 1000 or 100,000 people. If, however, the population of a state or city included a large number of adults, it had the capability of producing more divorces per 1000 people than a population with a large number of children. One investigator, for example, called Dallas, Texas, the "divorce capital of the world." He had established a divorce rate for 105 American cities by comparing the number of divorces with population numbers, apparently without allowing for the composition of each city's population.[8] The resulting statistics may have indicated only that the population of Dallas had a larger proportion of adults, and thus a greater ability to produce divorces, than other cities in the study.

Several other factors adversely affected the validity of the divorce rate. In earlier eras, marriages were of relatively short duration and characteristically ended with the death of one partner. But after World War II, most Americans lived longer than did their grandparents and great-grandparents had. Thus, the marriages they contracted in their teens and twenties could last for fifty or sixty years. Many unhappy marriages that previously ended with the death of one spouse were now ending in divorce, thus the high divorce rate that seemed to indicate that many more Americans divorced in the postwar decades than in previous generations. Such figures fail to take into account the relative longevity of marriages in different eras.[9]

In addition, the growing phenomenon of multiple divorce swelled the divorce rate. People who divorced more than once contributed generously to divorce statistics, with some logging as many as five or six divorces. Although it is true that each of these marriages ended in divorce, it was the same person or persons terminating one marriage after another. Because of the difficulties involved in tracing multiple divorcers in the records, it is virtually impossible to adjust the divorce rate accordingly.

The divorce rate is also inflated by the increasing difficulty of separating from, or deserting, a mate without taking legal action. A number of analysts believe that a higher proportion of separations and desertions ended in divorce after 1945 than before. In an age of Social Security numbers, drivers' licenses, retirement plans, and computerized records, it is increasingly difficult for spouses to go their own ways without engaging attorneys, appearing in court, and filing the necessary papers. Partings that would have once gone unrecorded are now divorce statistics, which expands the divorce incidence and makes it seem considerably higher.[10]

This is not to suggest that informal separations and desertions disappeared after 1945. On the contrary, scattered evidence shows that couples continued to part without benefit of divorce. In 1940, 3.1 million married people told census takers that they were living apart from their spouses. Although it is impossible to know whether illness, business, or preference underlay these separations, experts estimated that one-half of these people had parted permanently from their spouses. During ensuing decades, a variety of observers maintained that separation and desertion continued to be serious problems among all groups of Americans, especially among poor and uneducated African-Americans.[11]

Despite the discrepancies, divorce statistics did demonstrate several clear patterns. Whether the number of divorces was compared with number of marriages, total population, or other measures were used, the resulting divorce rate showed a marked increase between the mid-1940s and the late 1980s. It rose noticeably during the years following World War II, presumably due to the dissolution of wartime marriages and the end of marriages fractured by war-related stress. The rate settled on a plateau between 1955 and 1963, then resumed its ascent toward an all-time high by the late 1980s.[12]

As the divorce rate escalated, other trends became clear as well. Divorce rates in urban areas continued to be higher than those in rural districts. The western United States continued to have the highest divorce rate, while the South rose to second position and the Northeast dropped to third place.[13]

The growing availability of divorce statistics also resulted in an intensified study of divorce in specific groups. African Americans received considerable attention at least in part because they continued to divorce in greater numbers than members of other racial groups. One study argued that supposed female dominance was unrelated to the high divorce rate among African Americans; another theorized that more African-American families had adopted the egalitarian model than white families. Other analysts speculated that African Americans were divorce-prone because they exhibited several factors known to cause divorce: to marry young, have low rates of education and high rates of poverty, and experience marital stress as a result of racial discrimination.[14]

Other researchers focused on divorce among Mexican Americans, whose divorce rate was lower than that of whites and African Americans, probably as a result of Mexican Americans' adherence to Roman Catholicism. But investigators found that despite the church's influence, the rate of divorce among Mexican Americans was rising.[15]

Studies that examined divorce in religious groups demonstrated that divorce was also increasing among members of other conservative faiths. Mennonites, for example, divorced at a slower pace than Americans overall, but their divorce rate was rising. A comparison of American Mennonites and Canadian Mennonites showed that those living in the United States divorced at a rate four times greater than those living in Canada. Evidently, religious beliefs failed to shield American Mennonites from the factors that impelled Americans to divorce in large numbers.[16]

Still other investigators studied inter-group marriages, which are generally less stable than same-group marriages. Marriages involving partners of different races, ethnic backgrounds, and religious persuasions were found to experience a higher rate of breakdown than those involving spouses of the same race, ethnicity, or faith. As a case in point, during the early 1950s, cross marriages of members of different groups represented in the Hawaiian islands—Hawaiian, Chinese, Japanese, Caucasian, Filipino, Korean, Puerto Rican—ended in divorce more often than same group marriages.[17]

Other researchers analyzed statistics concerning women and children. After 1945, growing numbers of women employed outside the home sought divorces. In addition, divorce was increasing among families with children, a significant trend that continued throughout the 1980s.[18]

The increase in the numbers of children of divorce forced psychologists, home economists, counselors, and researchers to intensify their study of the effects of divorce on children. Beginning in the 1950s, a number of studies demonstrated that children adjusted to divorce within a few years, and generally fared better than children of high conflict, but intact, marriages.[19]

By the late 1980s, however, investigators had come to recognize the pitfalls in such research, including faulty methodology and inadequate or biased data. Researchers also revealed that a tremendous number of variables affected children's adjustment. These factors ranged from parent-child interaction to the attitudes of siblings to the type of counseling, if any, the child received. At the end of the decade, several investigators argued that divorce was a major trauma for children, a trauma that affected them, very often negatively, the rest of their lives.[20]

At the same time, anecdotal evidence indicated that it was possible for some children of divorce to handle it well. In 1990, for example, columnist Ellen Goodman described the wedding of two adult children of divorce. Both had "lost the parental umbrella at an early age." Although they grew up in shared custody, "they felt a loss of family that came with this surplus of families." Yet they both triumphed, found each other, and decided that they did not want their own children "to go through these same rites of passage."[21] But, of course, it was impossible to know whether these determined children of divorce would succeed in holding their own marriage together.

Clearly, one of the major difficulties in studying children of divorce is following them through their own life cycles. Another is assessing the impact of economic conditions, social class, ethnicity, race, and education on their adjustment patterns.[22] Despite the existence of a growing literature on children of divorce by the beginning of the 1990s, more study was indicated because of the many questions that remain unanswered and the steady increase in the numbers of children of divorce.

## No-Fault Divorce

As the divorce rate spiraled upward, more people became concerned about the effects of the divorce process on divorcing spouses and their children. Early in the twentieth century, several Americans began to support a radical plan to soften the process by replacing adversarial divorce with divorce by mutual consent. As early as 1915, one commentator prophesied that divorce by mutual consent was "likely to form one of the provisions of future divorce law." In 1923, novelist and essayist Katharine Fullerton Gerould had argued that if laws made it possible "to marry at sight," they ought to make it possible "to divorce on demand." And in 1927, Judge Ben B. Lindsey of Denver suggested that couples undertake trial marriages that could be terminated by mutual agreement.[23]

During the mid-1940s, Americans continued to discuss the possibility of replacing adversarial divorce with a non-punitive procedure.

In 1947, the New York Public Affairs Committee published a pamphlet, *Broken Homes,* criticizing adversarial divorce. It maintained that because most husbands and wives were driven apart by "internal tensions" in the marriage itself, both parties were "at fault." Two years later a legal authority defined divorce as the termination of an unworkable relationship, a termination that should consider the best interests of divorcing spouses and their children rather than punishing the "guilty" party.[24]

Other writers explored the form that mutual consent—essentially no-fault—divorce might take. In 1949, one legal specialist supported the American Bar Association's "contract theory" of divorce. A dissatisfied couple would present data about themselves and their children to a judge. During the next six months, the court would investigate the case, while the couple considered reconciliation. If the information submitted proved accurate and the couple rejected reconciliation, the judge would grant a divorce. This process was intended to bring an end to court battles "about 'causes'," "substitute honesty for hypocrisy," and result in humane, enforceable divorce decrees.[25]

In the meantime, a number of states by legislative enactment had begun to de-emphasize adversarial divorce procedures. In several states, statutes permitted couples to divorce after they had lived apart for a specified time; no wrongdoing need have occurred. By the mid-1960s, eighteen states, Puerto Rico, and the District of Columbia sanctioned living apart as a ground for divorce.[26]

Another attempt to offset the harsh effects of adversary divorce was the establishment of family courts; only judges trained in family law would rule on such issues as alimony and child custody. It was hoped that family courts would establish equitable financial settlements and prevent one parent from denying child-visitation rights to the other. After the first family court was established in Cincinnati, in 1941, the concept spread to Milwaukee, St. Louis, Omaha, Des Moines, Portland, Oregon, and Washington, D.C. In 1949, the states of Texas and Washington initiated similar reforms. The Los Angeles Children's Court of Conciliation, which tried to reconcile divorcing parents, was perhaps the most well-known of these schemes. By 1959, the Los Angeles Children's Court reportedly reconciled 43 percent of the alienated couples who entered counseling.[27]

Still, many people continued to advocate the elimination of adversary divorce. In 1966, historian Christopher Lasch argued that lenient, non-punitive divorce would protect the family rather than threaten its well-being. If mates could easily end destructive marriages, society would be left with "mature marriages" rather than non-functional ones. Two years later, anthropologist Margaret Mead advised Americans to view divorce as the termination, without recrimination, of a dead marriage. Divorce was, in her view, an opportunity for a divorced person to form "a better marriage, a true marriage" in the future.[28]

During the late 1960s, such legal specialists as New York University law professor Henry H. Foster, Jr., and attorney Doris Jonas Freed recommended the adoption of mutual consent, or no-fault, divorce. Foster and Freed argued that divorce law must strike a balance between public concern for the family and the need to end an individual marriage. Foster explained that 70 percent of divorce petitioners tried to minimize the adversarial nature of divorce proceedings through collusion or by charging cruelty rather than utilizing harsher grounds, especially adultery. Foster predicted that "substantial reforms" were in the offing and hoped that they would "sever the bonds of acrimony" and assist divorcing mates and their children in getting on with their lives.[29]

California was first to act on the idea of no-fault divorce. After studying various suggestions and the 1966 Report of the Governor's Commission on the Family, California legislators began to draft no-fault divorce statutes in 1967. In 1969, the legislature approved the Family Law Act. Governor Ronald Reagan signed it into law on September 5, 1969. The bill, which went into effect on January 1, 1970, replaced California's seven grounds for divorce with two no-fault provisions: irremediable breakdown of a marriage and incurable insanity. Petitioners had only to reside in California for six months before applying for a divorce. Judges could award alimony based on a spouse's need for support and the other spouse's ability to pay; they were to divide a couple's property equally. Judges were also to make child custody decisions. And the final decree was to be known as a dissolution, rather than a divorce. Legislators hoped that these provisions would end adversarial divorce and eliminate the need for one petitioner to present evidence blackening the character of the other.[30]

California's no-fault divorce statute marked the beginning of widespread changes in American divorce law. In 1971, Iowa was the first state to follow California in adopting no-fault divorce. By August 1977, only three states retained the adversary system of divorce: Illinois, Pennsylvania, and South Dakota. Fifteen states—Arizona, California, Colorado, Delaware, Florida, Iowa, Kentucky, Michigan, Minnesota, Missouri, Montana, Nebraska, Oregon, Washington, and Wisconsin—stipulated irretrievable breakdown of a marriage as the sole ground for divorce, while sixteen others—Alaska, Alabama, Connecticut, Georgia, Hawaii, Idaho, Indiana, Maine, Massachusetts, Mississippi, New Hampshire, North Dakota, Ohio, Texas, Tennessee, and Rhode Island—had added irretrievable breakdown to existing "fault" grounds.[31]

The winds of change would seem to have blown widely, yet a number of even more radical plans were being touted during the 1970s. Proponents of newer strategies pointed out that no-fault divorce was only a minor revision of the traditional system of divorce—it had only formalized what litigants and judges had been practicing through collusion and lenient application of the charge of cruelty. These reformers believed that the existing system of divorce should be scrapped and a new system put in place.

Two plans gained widespread support. One was contract marriage: a couple would marry by means of a contract that could be renewed or terminated after a specified number of years. Contract marriage would reverse existing divorce law because spouses would have to take action to continue their marriage rather than taking action to terminate it. The second plan was registration divorce, also known as divorce on demand. A divorce would be granted if one or both spouses registered a divorce request. The state's adjudicatory role in divorce would become administrative; officials would only have to record requested divorces.[32]

Apparently, a number of Americans agreed with journalist Shana Alexander that "divorce should be as painless as possible." She reasoned that "as long as people so often persist in getting married for all the wrong reasons, or before they have matured sufficiently to know what they're letting themselves in for or who they really are," they must have a convenient escape hatch. After all, divorcing spouses usually suffered enough "emotional trauma" without having to "endure a legal trauma as well."[33]

Still, no additional modifications occurred in American divorce. Further changes may have seemed unnecessary because experts and laypersons alike expected great results from no-fault divorce. As early as 1970, sociologist Jessie Bernard predicted that no-fault divorce would remove the last vestiges of stigma from divorce. She foresaw independence for women who were increasingly able to support themselves after divorce because of "spectacular" employment trends. And, drawing upon studies showing that children fared better after divorce than in a disintegrating marriage, she foresaw stable mental health for children of divorce. Bernard's optimism regarding no-fault divorce mirrored the hopes of millions of Americans.[34]

No-fault divorce did bring about a number of significant alterations in American divorce. It changed the concept of divorce from a punishment of an offending spouse to a "remedy for situations which are unavoidable and unendurable." And, although no-fault divorce law attempted to preserve marriages when possible, it provided relief to dissatisfied spouses on non-judgmental grounds.[35]

It also made collusion between divorcing spouses unnecessary because one no longer "sued" the other for divorce. As a result, the number of male petitioners increased considerably. Between 1966 and 1969, more than 78 percent of plaintiffs in California divorce suits were female. In 1974, this figure dropped to slightly below 68 percent.[36]

No-fault divorce also led to less costly divorce because a no-fault action seldom involved high court, attorney, and other costs. A 1977 U.S. Supreme Court ruling in *Bates v. State Bar of Arizona* intensified this effect by permitting attorneys to advertise their services. Soon, American lawyers advertised easy, inexpensive divorces more frequently than any other legal service. According to these advertisements, divorces could be obtained for the bargain price of $50 to $100.[37]

Unfortunately, no-fault divorce also had some destructive effects. A few commentators pointed out the flaws of no-fault divorce shortly after its adoption in California and Iowa. As early as 1973, the author of a women's divorce manual noted that although no-fault divorce had been "hailed as the welcome end" to finding one party guilty and the other "a paragon of virtue," it might also bring "irremediable breakdown of the marriage institution." When an irresponsible husband wanted to leave his marriage lightly, how could a newly vulnerable wife protect herself and her marriage? "Doesn't it seem fair that if one party seeks to preserve the marriage, the other should then be obliged to prove that because of her misconduct, he's won the right to leave her?"[38]

Two years later, a *male* commentator pointed to the same crucial weakness in the no-fault system. No-fault divorce undermines "the bargaining position of a dependent spouse" because the court has the power to determine "support and maintenance awards on the basis of demonstrated need and merit" rather than on a mate's guilt. He stated that this was "a small price to pay for an honest system of determination." He prophetically concluded that this practice "could be unfair in its impact in a particular case."[39]

By the mid-1980s, it was apparent that Americans' honeymoon with the concept of no-fault divorce was over. For one thing, no-fault divorce had given migratory divorce a new twist. The *Wall Street Journal* explained that divorce-seekers no longer shopped for a jurisdiction with a short residency requirement, but looked for a state where they could obtain the most favorable property, alimony, and child custody awards. The author speculated that California statutes tended to favor the spouse with less property than the other because of its community property provisions requiring that all property be split equally. Of course, the complication of such a ploy was that there could be a jurisdictional battle between one spouse petitioning for a divorce and *favorable* provisions in one state, while the other spouse tried to achieve *favorable* ends in another.[40]

A second, far more serious, result was the contribution of no-fault divorce to the feminization of poverty in the United States. In 1986, an article in the *New York Times Book Review* proclaimed that "no-fault" was "no fair." Its author described a number of books that documented in chilling detail the destructive effects of no-fault divorce statutes on women's economic standing. They also exposed a growing trend toward women's loss of custody of young children, wives' lack of redress for a husband's adultery and defection from the marriage, and the failure of judges to offset women's low earnings with proportional property settlements.[41]

What had happened to the brave vision of the late 1960s and early 1970s? How had legislators adopted no-fault divorce without any searching analysis of what it would mean for women and children?[42]

During the 1980s, experts tried to answer these questions. Sociologist Lenore Weitzman explained that creators of no-fault divorce laws had intended to establish norms for property settlements and alimony based on the concept of wives as full economic partners and failed to realize that no-fault provisions would have dire economic consequences for many divorcing women. In practice, few wives earned as much as their husbands. If women were full-time wives and mothers, they seldom earned any cash income at all, but instead, sacrificed their own education, training, and job experience to enhance their families' well-being and their husbands' earning power. Thus, even if they had worked outside their homes, most divorcing women lacked the financial wherewithal to support themselves and their children adequately.[43]

Even when a woman's husband committed adultery and left her for another woman, she received an equal share of property but little or no alimony, although she was free of wrong-doing. Although he severed the marital bond, she suffered because she had to support herself without sufficient skills or experience. Alimony was usually temporary due to the assumption that she would quickly become self-sufficient. If she received custody of the couple's children, her woes were often increased. Although judges ordered fathers to pay child support, the amount was usually inadequate. In addition, the default rate was high and support orders frequently went unenforced. At the same time, public assistance was not prepared to support all needy women and their children. Nor could it provide enough job training and child care to pull all destitute women out of their financial morass. The result, according to Weitzman, was that many a divorced wife's income dropped sharply after divorce, while her former husband became significantly better off.[44]

During the 1980s, examples of troubled no-fault divorces abounded. In a randomly selected sample of sixty divorced women of various ages, social classes, races, ethnic backgrounds, and regions of the country who had divorced under no-fault provisions collected for this study, forty-six respondents related minor to serious problems in supporting themselves and their children after divorce. These women were hampered by inadequate education, and a lack of skills and paid work experience. Forty of the forty-six had one or more children, which increased their expenses significantly.[45]

One interviewee was a Los Angeles mother of three who had worked as a legal secretary before her marriage. When she married an attorney and became pregnant, he urged her to leave her job. Nine years later, they parted by mutual consent and agreed that she would have custody of the children who were in the "tender years" age range. He agreed to pay child support, which proved to be too little to offset her high expenses. Because he was about to marry a woman with children, he had no extra money to give her. With the financial and child-care aid of her parents, this Los Angeles woman returned to college. At the time of the interview, she was balancing class work with the care of a home and three young children. Whether she would survive this burden long enough to become financially independent remained to be seen.[46]

Another happier story concerns a St. Louis woman who found only a low-paid job as a receptionist after her divorce. With the aid of her employer, who paid part of her tuition and her daughter's child-care expenses, she earned a B. A. degree in business administration. After many years of working full-time and attending school part-time, she moved into a more responsible, well-paid position with her company.[47]

Only fourteen of the women in the sample fared well from the outset with their no-fault divorces. Five cases involved children. In three of the cases, the former husband held a well-paying job which enabled him to help his former wife and their children. In two of the cases, former wives held well-paying jobs. Among the divorced couples who had no children, one woman's former husband helped her return to school. Two other women were able to return to former jobs, while five already held reasonably good jobs. One of the these was a Pasadena woman who was a university professor whose former husband was also a university professor. When they divorced after five years, they split their property and their debts equally. The former wife felt that both were satisfied with the settlement.[48]

This sample is clearly too limited to justify sweeping conclusions, but it does demonstrate that no-fault divorce worked against many divorcing women during the 1980s, especially if they possessed little education, job skills, and work experience and received custody of several minor children. The interviewees who described their economic plight had few resources available to them other than their families and Aid to Dependent Children. Several remarked that if they remarried, they would continue to work because they wished to avoid economic hardship in the event of another divorce. Their viewpoint seemed justified because, according to this sample, of the women who survived no-fault divorce with ease, most held good jobs.

### Singlehood and Remarriage

Divorced Americans also had to tackle a multitude of other problems, including how to integrate themselves into the growing world of single people. As the divorced population grew in the United States, it swelled the number of one-person households and of families headed by single parents. By 1970, 17.1 percent of all American households consisted of one person; by 1985, this figure had jumped to 23 percent. During this period, families headed by single people more than doubled.[49]

The growing number of divorced people, especially single parents, faced a bewildering variety of new situations. By the 1980s, the ranks of custodial parents included a growing number of fathers who obtained full-time or joint custody of their children. These men, who resented the customary practice of awarding child custody to mothers, wanted to participate in raising their children rather than lose contact with them as so many divorced fathers did. In 1985, Census Bureau figures showed that more than 893,000 fathers headed single-parent families (including widowers and single men who adopted children) and were rearing more than one million children. Their concerns—along with those of non-custodial mothers now required to make child support payments—began attracting widespread media attention.[50]

New services began to spring up to meet the needs of single parents. Parents Without Partners was among the first to give support to divorced (and widowed) parents and their children. Organized in 1957, PWP reached out to single parents and their children through discussion groups, speakers, recreational and social activities, and a magazine called *The Single Parent*.[51]

By the 1980s, it was becoming apparent that many divorced people had an important problem: finding another mate. Media advice proliferated. One counselor suggested that "middle-age" dating should include breakfasts, shopping excursions, and time at home to foster reality and intimacy between a couple. At the same time, dating services began to tap the huge market of divorced people. Some dating services were computerized and offered a videotaped interview of a client's possible date. In addition, such national organizations as Singles in Agriculture counted many divorced people among their members. Other divorced people ran "singles ads" in local newspapers; if accompanied by a photograph, these could cost as much as $500 in New York City newspapers.[52]

Once they located a potential mate, many divorced people preferred to live with the person as a form of trial marriage rather than immediately marrying. In 1988, the Census Bureau reported that the number of unmarried co-habiting couples—including never married, divorced, and widowed people—had reached 2.3 million, including 700,000 couples who were raising children. Another 1.5 million same-sex couples were living together, including 92,000 with children.[53]

Occasionally, these unions too led to disputes concerning property and palimony. Although some co-habiting couples negotiated agreements and contracts, the courts found them difficult to interpret. In 1979, the decision in *Marvin v. Marvin* established the principle that courts should enforce explicit contracts between non-marital partners despite the contention that such contracts opposed public policy encouraging couples to marry. In the absence of a contract, the court was to investigate the relationship to determine if an implied contract or joint venture existed. If so, the court was justified in awarding palimony (monetary payments similar to alimony) to one of the partners.[54] In practice, however, palimony awards were small and infrequent.

Increasingly, psychologists and counselors cautioned co-habiting couples that living together was an invalid test for marriage. They argued that partners were usually well aware of the trial status of their relationship and thus inclined to maintain good behavior. In addition, co-habiting men and women tended to accept traits in partners that they would find intolerable over the long run. During the 1980s, studies confirmed that couples who lived together before marriage tended to get divorced sooner than couples who had not. In 1989, an University of Wisconsin study revealed that 38 percent of couples who had lived together prior to marriage divorced within ten years; 27 percent of couples who had not lived together did so.[55]

Those who remarried and left the problems of singlehood behind soon confronted a new set of hurdles, especially if they established a blended family—a family with one or two previously married parents and children from previous marriages and the present marriage. Beginning in 1972, over one million children a year witnessed the dissolution of their biological families by divorce. During the 1980s, one of seven children lived in a family in which one parent had remarried and at least one child was from a former marriage.[56]

Approximately half of all divorced mothers remarried within two years of their divorces during the 1980s. A woman's decision to remarry was often encouraged by financial need: only 15 percent of divorced mothers received alimony, and it was usually temporary and inadequate. Also, former wives frequently did not receive court-ordered child support payments: only 40 percent of fathers charged to pay support did so during the first year after a divorce; by the tenth year this dropped to 13 percent.[57]

Blended family situations were often highly stressful for everyone involved. Yet of children interviewed about their families a majority said that they preferred living in a blended rather than a single-parent family. The spouses in blended families were often less satisfied than the children, however, and divorced again. During the 1980s, 17 percent of remarriages that involved stepchildren ended in divorce within three years, while 10 percent without stepchildren ended in divorce.[58]

The problems of blended families ranged from financial disputes, to child custody fights, to endemic confusion concerning complex kin relationships. Although the last difficulty often seemed inconsequential to people about to remarry, it frequently exacerbated all other problems faced by blended family members.

Because women's traditional work of maintaining kin relationships by telephoning relatives, sending cards and gifts, planning holidays, and celebrating birthdays multiplied in a blended family, remarried women often discovered that it was impossible to keep up the kin work of both their previous and present families. Other women found themselves maintaining a "divorce chain," a pseudo-kinship network of former and current spouses, and the respective relatives.[59]

Incredible complications often resulted. One young woman described her family get-togethers as a "morass": "my 90-year-old grandmother" does not know how to introduce my father's second ex-wife to "my step-sister's 5-year-old daughter (who can't quite work out who I am)." This family was caught in the whirlpool of what sociologists termed "the blended-family gathering." After floundering in the middle of these events, the young women concluded that little blending occurred: "it's more like grinding or curdling."[60]

As the decade of the 1980s progressed, Americans directed queries about their blended families to newspaper advice columnists. An Alabama woman wrote to "Dear Abby" concerning her mother and her stepfather "Ralph" who had recently divorced. Ralph had remarried and had a baby boy named "Michael." She wrote: "Is Michael related to me? If yes, what is he to me? Is Ralph still my stepfather? I have a half-brother, 'Tommy' from the marriage of Ralph and my mother. Ralph's new baby, Michael, would, I believe, be Tommy's half-brother, right?" She also asked if her three-year-old son was related to Michael, and if her mother's new husband was her stepfather and her son's step-grandfather. Remarkably, Abby sorted out this muddle.[61]

Interviews with fifty-three remarried husbands and wives conducted for this study revealed similar complexities—and a high number of divorces. A case in point are Floridians "Bill" and "Alice" who married in 1970. He had been married twice before and had a child from the second marriage. Bill paid support for the child and had visitation rights for one month a year. Alice had been married once before, had one child in her custody, and her former husband neither paid support nor visited the child. Bill and Alice also had two children of their own. They divorced in 1978, then reconciled and remarried in 1979. In 1983, they, her child, and their two children attended Bill's son's wedding. They divorced again in 1988.[62]

Twenty-three of the blended family couples interviewed were still married. One of these is "Phil" and "Sarah" who married in 1976 in Philadelphia. Phil had three married children who each had a child. Sarah had joint custody of two minor children. Phil and Sarah had one child of their own. Still married in 1987, they were parents of young children and grandparents at the same time. Another example is a New York couple who married in 1971. It was her first marriage and his third. He had one son for whom he paid support and visited several times during each year. They had two children, then divorced in 1976. They remarried in 1979 and were still together in 1987. By then, his son was grown, but had not yet married.[63]

These interviews with remarried men and women expose some of the difficulties remarried spouses and their children faced. Remarriages crumbled under the weight of responsibilities to children and former spouses, financial and property concerns, differing habits and outlooks, two jobs or careers, scarred emotions, and interference by former spouses, relatives, friends, and co-workers. Given these and other problems confronting remarried couples, their high rate of divorce is unsurprising.

Despite the problems involved in remarriage, during the early 1980s more than three-fourths of divorced people contracted marriages to new spouses. Between 1970 and 1983, the number of remarriages of divorced women and men increased by 82 percent. In 1983, 33 percent of white brides had been previously married and divorced, while 24 percent of African-American brides had been married and divorced. During this fourteen-year period, a total of 8.2 million previously divorced women and 8.7 million previously divorced men remarried.[64]

The more frequently a person divorced, the shorter the projected duration of his or her next marriage. In 1983, the average duration of marriage for all divorcing couples was 9.6 years; for once-married women and men the duration was 10.8 years; for twice-married it was 7.0 years; and for thrice-married it was 4.9 for women and 5.1 for men.[65]

Yet hope springs eternal; many divorced people remarry immediately. In Reno and Las Vegas, wedding chapels welcome the newly divorced without question, and one chapel provides a bottle of champagne and a wedding cake to couples who respond to an advertisement in local newspapers.[66]

In 1985, an article in *Newsweek* described remarriage as a "drug-of-choice" for people "unable to tolerate the aftershocks of a failed marriage or the prospect of a lonely life ahead." Consequently, divorce lawyers often had repeat clients who divorced and remarried regularly. Sill, the article noted, most people about to remarry believed their new marriages would last a lifetime. Bridal consultants reported that brides entering a third, fourth, or fifth marriage glowed with happiness and optimism, and often insisted on a white wedding gown.[67]

Many of the divorced women and men mentioned earlier in this study also remarked that they felt extremely hopeful at the beginning of a remarriage, and they were shocked and disappointed when it ended. One interviewee who had been divorced four times remarked, "Each time I marry, my hopes and emotional investment are higher because I think that with maturity I will surely get it right this time. . . . When the marriage flops, it throws me into a worse depression than the last one."[68]

In an attempt to avoid failed remarriages, many people negotiate pre-marital agreements. Conventional wisdom of the 1970s and 1980s instructed remarrying people to be practical, but to avoid destroying the romantic glow tinging their relationships. Because some divorced people brought substantial financial assets as well as obligations to marriage, legal experts advised couples to settle before they married questions of separate and joint property, the division of assets in the event of divorce or death, and children's claims on property. Supposedly, such agreements would take care of situations like the one faced by a distressed wife who worked full-time while her husband's former wife worked part-time. By her own admission, she "used to really blow up" when she "wanted to buy something like a dishwasher, and he said we couldn't afford it."[69]

Serial marriage and divorce appeared to be on the increase during the 1980s, but the extent was difficult to document. Many people lied to clerks, officials, and census-takers regarding such personal information as age, marital status, and whether they were divorced or widowed. Although statisticians knew that far more women reported themselves as widows than could have been possible according to other census data, they failed to find a way to correct the inaccurate statistics that resulted.[70]

To judge from media coverage, the phenomenon of multiple divorce was growing. Certainly, Americans had a fascination with people who frequently married and divorced, especially those who were Hollywood stars. In 1959, a *McCall's Magazine* article reported the marital escapades of Hollywood figures and admonished them for their actions. In what purported to be the first chart of Hollywood divorces ever compiled, the following information was revealed: in 1934, Gloria Swanson divorced her fourth husband; in 1947, Rita Hayworth and Orson Wells divorced; in 1949, Dick Haymes and Joanne Dru divorced; in 1955 Dick Haymes and Rita Hayworth divorced; in 1957, Joanne Dru and John Ireland divorced; in 1950, Arline Judge divorced her sixth husband; in 1956 Artie Shaw divorced his seventh wife; and in 1957, Elizabeth Taylor divorced Michael Wilding and wed Mike Todd.[71]

In subsequent years, such stars as Zsa Zsa Gabor, Mickey Rooney, and Elizabeth Taylor contributed heavily to Hollywood's divorce and remarriage rate. In 1988, an *USA Weekend* article exposed a number of other well-known personalities who had married and divorced several times. To the query "how many marriages?" it responded: Muhammad Ali, 4; Glen Campbell, 4; Johnny Carson, 4; Joan Collins, 4; Dick Clark, 3; Doris Day, 4; Rita Hayworth, 5; Liza Minnelli, 3; Mary Tyler Moore, 3; Richard Pryor, 5; Jason Robards, 4; Kenny Rogers, 4; Telly Savalas, 3; George C. Scott, 5; Frank Sinatra, 4; Jane Wyman, 5; and Tammy Wynette, 5.[72] Unlike the 1959 article, this one encouraged people to take a chance on remarriage and assured them that it was acceptable for them to marry again—and again.

## Helping Oneself

Between the mid-1940s and the late 1980s, while the number of divorced Americans edged steadily upward, their problems seemed to snowball. Increasingly, divorced men and women sought advice from marriage counselors, support groups, and reading matter. Consequently, divorce self-help literature grew from a few books usually shelved in bookstores under "Marriage and the Family" to a sometimes bewildering assortment of books arranged under the headings of "Self-Help" or "Psychology." In addition, their counsel increasingly was supplemented by newspaper and magazine articles. By the 1980s, people contemplating divorce as well as divorced people about to undertake dating, co-habitation, or remarriage could easily find any number of manuals and articles to guide them through the experience and offer them solace along the way.

A survey of the attitudes of representative self-help authors between 1945 and 1990 discloses the changing problems of divorced people during these years. It also reveals the transformation of divorce from a mark of failure to an opportunity for growth, from an occasional occurrence to a common phenomenon.

During the 1940s and 1950s, advice concerning divorce was scarce. A few authors of marriage manuals included small segments of guidance to people who were thinking about divorce, or who were involved in divorces. These authors drew heavily upon prewar and media views of divorce in formulating their counsel and dire warnings.

One typical 1940s marriage manual decried the "evils" that divorce contributed to American society. Its author maintained that marriage was a lifetime undertaking, and argued that divergent state laws led to abuses that in turn "diminished the regard for the sanctity of marriage." For those readers who insisted on obtaining a divorce, he cautioned them that out-of-state and out-of-country divorce could lead to battles between a first and second wife concerning a husband's estate, and might even result in bigamy charges against the husband. The author also urged readers to avoid contracting a remarriage before a divorce decree was final, especially in states that required waiting periods before remarriage. This advice was probably little used by the average divorce-seeker, who was planning neither a Reno-style divorce nor immediate remarriage.[73]

A few guidebooks devoted entirely to divorce were also available. Although these divorce manuals contained more thorough instructions than marriage manuals, they offered an equally dismal picture of divorce. One, titled the *ABC of Divorce,* first appeared in serial form in *McCall's Magazine* and was aimed at female readers. "You will not find this book entertaining because there is nothing entertaining about divorce," its authors warned. "At best, it is a brutal and bewildering experience." Calling divorce "the nation's Number One legal and social problem," the authors predicted, accurately as it turned out, that if the pro-divorce trend continued, in twenty years one of every two marriages would end in divorce. The *ABC of Divorce* also drew a dire picture of life after divorce. A divorcée could expect to "be alone in a lonely new world," be shunned by family and friends, and perhaps spend the rest of her life "without companionship of some kind."[74]

This dim view of post-divorce life was closely linked to the prevailing belief that women could find fulfillment only as wives and mothers. The "back-to-the-home" movement for women was in full force during the 1940s and 1950s. Dr. Benjamin Spock, whose *Baby and Child Care* reached one million readers every year between 1945 and 1960, preached that good mothers neither worked outside the home nor were away from their children for long periods of time. Spock advised mothers who were thinking of finding employment to seek psychological counseling instead and to consider seriously the potential delinquency of their children.[75]

The primacy and importance of women's domestic role were also reiterated in popular magazines during the 1940s and 1950s. In 1956, a special issue of *Life* summarized prevailing attitudes. The magazine's editors chose a thirty-two-year-old mother of four as the ideal American woman. She was a wife, hostess, volunteer, and "home manager" as well as a conscientious mother who spent "lots of time with her children, helping with their homework and listening to their stories or problems."[76]

Also in this special issue of *Life,* anthropologist Margaret Mead explained that American women were expected to be primarily wives and mothers. They were to "earn a living, manage money, hold a full-time job, run a farm, keep a store, teach a class or even govern a state," but only "if they have to." Both husbands and wives hoped "that none of these things will actually happen." Even if a wife worked, it was "thought of as supplementary," to be discontinued when she had children or her family no longer needed the money. Mead concluded, "the American woman lives in a world where she can be satisfied with nothing but marriage."[77]

Other *Life* contributors bemoaned women's eagerness to revise their roles. One derided the "shrill, ridiculous war" that women waged "over the dead issue of feminism," while another attributed the rising divorce rate to wives who fell short of being truly feminine and husbands who failed to act sufficiently "male."[78]

Some Americans challenged these views. One of these was sex researcher Alfred Kinsey, who recorded the sexual histories of individual Americans. Kinsey's reports of his findings in *Sexual Behavior in the Human Male* (1948) and in *Sexual Behavior in the Human Female* (1953) challenged customary conceptions. The Kinsey Reports, as they were known, revealed a wide range of sexual activity among women and men. They also documented a definite liberalization of sexual behavior among American women born after 1900. Kinsey also expressed support for women's sexual rights and assaulted a long-held belief by labeling a vaginal orgasm a "biologic impossibility."[79]

Traditional conceptions of women soon came under attack from other quarters as well. In 1961, President John Kennedy created the President's Commission on the Status of Women. Under the leadership of Eleanor Roosevelt, the commission recommended among other things, the establishment of government-funded child-care services for working mothers as well as equal employment opportunities for women. Two years later, Betty Friedan's *Feminine Mystique*, by exposing the frustration that many felt concerning domestic life, supported the idea that women wanted to work.[80]

During the early 1960s, divorce self-help literature began to reflect alterations in American women's roles and attitudes. Although writers and other commentators had linked women's 'emancipation' to the rising divorce rate since the 1920s, authors of divorce guidebooks now sharply indicted working women as culprits. In 1963, one writer declared that the high divorce rate was a direct result of "the emancipation of women—economic, legal, sexual, and intellectual." Working wives had become so aggressive "that the war between the sexes" had accelerated to a "overt and real" stage rather than being "disguised and subtle as before." Many wives now demanded more from marriage than their husbands could give. Working wives also constituted a particular "hazard" for marriage because of their inability successfully to combine jobs with care of home, husband, and children.[81]

Still, this author proffered a hopeful picture of divorce to female readers. The same aggressiveness that propelled women out of marriage was the very characteristic that would prevent them from withering away from loneliness after divorce. In contrast to the 1940s image of a divorced woman, books of the 1960s described a post-divorce life in which a divorced woman would make the necessary adjustments, find new friends, and perhaps even establish an amicable relationship with her former husband. She might remarry or she might choose to remain unwed. If the latter, she could still lead a "full, satisfying life" because she might find, like so many other unmarried people, that career rewards could substitute for the "satisfactions of marriage."[82]

This relatively optimistic view of post-divorce life expanded during the 1960s. In 1966, Morton Hunt described divorced Americans as members of a "large, half-hidden society . . . a veritable World of the Formerly Married," which had its own rules, opportunities, and mechanisms to bring people together. Should divorced people feel like failures, he asked? He challenged a widespread assumption of the 1960s that divorce was indeed a mark of failure. But Hunt urged divorced Americans to think of divorce as a "painful but necessary" decision: an "act of courage" and "an affirmation of one's belief in the value and the possibility of happy marriage."[83]

During the early 1970s, conventional wisdom regarding marriage and divorce expanded in yet another direction. Participants in a rapidly growing feminist movement were espousing a new form of marriage based on equality and communication. As a result, unlike the self-help books of the 1940s and 1950s that had portrayed women primarily as docile wives and mothers, those 1970s books argued that women were independent beings who expected, and deserved, reciprocity in their martial relationships.

Readers were assured they could avoid divorce if they developed reciprocity in their marriages. As early as 1970, sex researchers William Masters and Virginia Johnson suggested that mates could settle sexual and other conflicts if they respected "one another as independent and equal human beings" and would "talk realistically" in a spirit of reciprocity.[84]

The theme of reciprocity in marriage, which echoed an earlier call by novelist T. S. Arthur and reformer Robert Dale Owen, appeared in a host of other books during the 1970s. One of these was *Creative Intimacy,* published in 1975. Its author stated that lasting marriages were based on equality and mutual commitment. In that same year, former ambassador Clare Boothe Luce chided the media for its "round-the-clock" images of woman as the "child-wife." She, too, believed that only partners who viewed each other as equals could achieve a happy marriage. And, in 1976, a husband and wife team recommended that a wife retain her family name as a symbol of her equality with her husband.[85]

Still other books described methods of achieving reciprocity in marriage. These manuals included a variety of approaches, from Nena and George O'Neill's *Open Marriage* (1973), which prescribed that mates needed to spend time with other people and perhaps even establish sexual liaisons to John Powell's *The Secret of Staying in Love,* which argued that married couples needed to improve communication. Other manuals were the work of George Bach and a variety of co-authors who wanted mates to learn how to fight fairly, achieve genuine intimacy, and be aggressive in a creative way. Typically, self-help authors implied that if one adopted their philosophy or learned their technique, one could avoid divorce. From self-counseling and biofeedback to no-fault marriage, self-help manuals offered a way to avoid becoming a divorce statistic.[86]

If, however, a person did face a divorce, then there were numerous guidebooks to guide him or her through the process. Unlike books of the 1940s and 1950s, manuals published during the 1970s and 1980s carefully shunned the word "failure" and emphasized a practical approach to divorce. They provided detailed information to readers facing divorce in the modern age. Because no-fault, community property, and other recent developments had complicated divorce laws, self-help authors explained the intricacies of divorce. A 1983 guide answered such questions as how to find a lawyer, how to tell children about a divorce, and what to do about safe-deposit boxes. It also included checklists, time schedules, and worksheets to help readers prepare credit histories, establish child visitation schedules, and be effective witnesses.[87]

A number of these books addressed themselves specifically to female readers. One was a manual designed by counselors at the Philadelphia Women's Center. In 1971, staff members responded to requests for advice by founding a resource program called Women in Transition. This program attempted to provide "survival skills and emotional support to women," especially African American, Puerto Rican, and low-income women who were "experiencing separation, divorce, and/or single parenthood." In 1975, the program produced a national edition of its guidebook, *Women's Survival Manual: A Feminist Handbook on Separation and Divorce.*[88]

Other programs offering self-help advice and literature sprang up across the United States during the 1980s. One called the Divorce Centers had a branch located in the medium-sized industrial city of Waterloo, Iowa. As a private consumer group that assisted divorce-seekers in preparing their own uncontested divorce paperwork and acting as their own attorneys, the Center's objective was to enable its clients to dispose of simply and inexpensively the legal matters related to divorce. According to the director of the Center, it offered a "new age approach to divorce."[89]

But many divorce-seekers needed emotional rather than legal advice. To meet this need, a new and distinctive type of self-help book appeared on bookstore shelves during the 1970s and 1980s. These promised to sustain their readers through the emotional aspects of divorce and to help them recover from the loss of people they had once loved, or perhaps still loved. They were a far cry from books of the 1940s that counseled divorced women to expect loneliness, and from those of the 1950s that advised readers to substitute the satisfactions of work for those of marriage. Instead, these books hoped to help their readers achieve personal growth through divorce and to find happiness as divorced people.

Written by such practitioners as marriage counselors and psychotherapists, these 1970s and 1980s self-help books set forth concepts arrived at through counseling patients and workshops. One early entry was counselor Mel Kranztler's *Creative Divorce* (1973), in which he bared the facts of his own divorce and asserted that potential for growth underlay every divorce. A few years later, *How to Survive the Loss of a Love* took another tack by interspersing fifty-eight practical suggestions of "things to do when there is nothing to be done" with bits of poetry.[90]

Others offered advice regarding the emotional aspects of relationships. Psychotherapist Howard Halpern's *How to Break Your Addiction to a Person*, published in 1982, described how "attachment hunger" formed the basis of one person's dependency on another. Halpern also tried to help readers decide if, how, and when to end a relationship, and how to proceed after doing so. The following year, another guidebook added a new dimension to this literature. Rather than helping a person leave a partner, it contained "sure-fire strategies for getting back together and making it last."[91]

Self-help manuals were, increasingly, supplemented by magazine and newspaper articles. In 1979, *Ebony* advised its readers to survive "one of the worst times" of their lives by seeking counseling, avoiding jumping into a new relationship, and forgiving the former spouse. In 1983, *Ebony* gave additional suggestions and lamented the lack of helping services for divorcing African American women.[92]

The themes appeared in hundreds of variations. In 1986, a *New Woman* essay took a hard line by advising readers to engage a good lawyer and a certified public accountant, and to prepare evidence in case a divorce action resulted in a trial. In that same year, *Psychology Today* reported a study that analyzed the benefits of establishing "Perfect Pals," "Cooperative Colleagues," "Angry Associates," or "Fiery Foes" relationships with former spouses.[93]

Despite this massive literature, divorced people about to enter the complicated world of single people during the 1970s and 1980s still faced a plethora of problems they often felt ill-equipped to handle, and many turned to a growing literature aimed at never-married, divorced, and widowed men and women. In 1974, *The Challenge of Being Single* assured readers that it was "okay to be single," counseled them about aspects of the "single experience," and touted the numerous advantages of "freedom." In 1981, a guide titled *Living Alone and Liking It* advised readers how to combat loneliness, entertain oneself, cook, and take care of oneself. Proclaiming that living alone could be a positive experience, it also promised "in living alone you are endowed with a most significant kind of freedom: the freedom to manage your time and your life in any way you choose."[94]

During the 1970s and 1980s, yet another type of self-help literature appeared. Directed to women, it told never-married, divorced, and widowed women how to handle, understand, and change men. The appearance of books of this sort coincided with a climb in the divorce rate, the entry of huge numbers of women into the labor force, the growth of the feminist movement, and the implementation of affirmative action policies. It seems likely that millions of newly independent, and perhaps newly divorced, women bought these books, which promised to help relieve them of an emotional and financial dependency that was less and less useful in their world.

This particular category of the genre covered an unbelievably wide range of topics. Volumes titled *Men Are Just Desserts* and *A Hero Is More*

*Than Just a Sandwich* urged women to become independent, to develop lives of their own if they wished to enrich the time they shared with men, and to learn how to stop choosing the wrong partners. Others titled *The Peter Pan Syndrome* and *The Wendy Dilemma* explored men's little-boy-like dependency on women and the need of numerous women to continue serving as dependency figures to men. Still others, *Some Men Are More Perfect Than Others* and *How to Raise Your Man*, analyzed the problems of a "new-style" woman in love with an "old-style" man, while *Why Do I Think I Am Nothing With a Man?* and *How to Ask a Man* encouraged women to revise some of their traditional ideas about female and male roles. A book titled *Women Who Love Too Much* examined women's co-dependency in love relationships, while *Men Who Hate Women & The Women Who Love Them* discussed men's emotional abuse of women. And *Smart Women, Foolish Choices* advised women how to find the "right men" and avoid the "wrong ones."[95]

Because most of this woman-oriented literature was written by counselors and therapists, it was often disparaged as "pop psychology." Some critics charged that its authors were exploiting the liberation, and new purchasing power, of women. But its authors seemed to believe that they were sincerely trying to answer a need of contemporary American women. Evidently, neither writers nor readers foresaw that the tone and volume of this women's self-help literature would border on what was termed "male bashing" during the late 1980s.

No comparable literature appeared on the bookstands for men during these years. Men were left unguided in their attempts to handle, understand, and change women. Nor did books attempt to tell men how to pick the right woman out of many wrong ones or analyze women's emotional abuse of men. Rather, the few male-oriented books published during the 1980s tried to make sense of the feminist movement and modern views of women from a male point of view.

In his trilogy—*The Hazards of Being Male, The New Male,* and *The New Male Female Relationship*—psychologist Herb Goldberg examined masculinity, defined an emerging male consciousness, and maintained that gender conditioning poisoned man-woman relationships. To him, divorce was a volatile situation because it "triggers the surfacing of dormant, underlying feelings that have been kept repressed at the inevitable result of the romantic, illusion-filled approach to courtship and marriage in our culture." Because Goldberg feared that women and men would "cement themselves into hardened postures of mutual antagonism and alienation," he called for far more than had earlier self-help authors: he wanted man-woman relationships based upon "authentic friendship, companionship, and sexuality."[96]

Educator and reformer Warren Farrell took a similar stance in *The Liberated Man* and *Why Men Are the Way They Are*. Farrell probed men's and women's beliefs about themselves and each other. He also described a "relationships revolution." According to Farrell, "now there are blended families and multiple orgasms; dual careers and shared housework, hyphenated names and communications seminars." Today, a woman does more than "give a man sex;" she expects "multiple orgasms, simultaneous orgasms, sensitivity, and sensuality." Consequently, Farrell concluded, training in traditional sex roles was tantamount to divorce training, for it created women and men who were opposites in their interests, home and child-care duties, and types of employment. Because "opposites attract" but "can't live together," they would eventually join the ranks of the divorced.[97]

By the late 1980s, then, not only was self-help literature far different from that of the 1940s, but it reflected a different society. Its sheer volume indicated how widespread and accepted divorce had become in the United States. Its diversity disclosed the numerous problems divorced men and women faced. And its emphasis on personal growth made it clear that divorce was no longer a mark of failure. Rather, divorce was a matter to be managed much like life's other significant milestones—marriage, giving birth, and death.

In 1986, an article in the *Des Moines Register* explained the new etiquette of divorce. A month later, a Chicago restaurateur offered to ease customers through divorce with a free meal if they came to his establishment on the day they received their decrees. Shortly thereafter, Hallmark Cards, Inc., began to market red and white, heart-sprinkled coffee mugs bearing the maxim, "Tis better to have loved and lost than to be stuck with a real loser for the rest of your entire, miserable existence!"[98]

Obviously, divorce had a new image in the United States. It was no longer a matter of shame or a closely guarded family secret. Instead, a divorce was usually a matter of general knowledge, sympathy, and discussion—and sometimes an occasion for humor.

Yet, despite its widespread acceptance in American society, divorce was far from problem-free. No-fault divorce, which marked its twentieth anniversary on January 1, 1990, had failed to fulfill its promise as a sweeping reform.[99] Marriage and divorce were fraught with danger because such crucial issues as spousal support, property, child custody, and remarriage continued to require attention and reform.

# Epilogue

A fable engraved in the main gate leading to the city of Agra, India, reported that during the first year of King Julief's reign, magistrates divorced two thousand couples. When he heard about these divorces, the King was so outraged that he abolished divorce. During the year following his proclamation, the number of marriages dropped by 3000, while the number of adulteries rose by 7000. Officials estimated that embattled couples destroyed three million rupees' worth of furniture. In addition, three hundred wives were burned alive for poisoning their husbands, and seventy-five men were executed for murdering their wives. Upon learning of the aftermath of his decree, the emperor re-established divorce.[1]

Hundreds of years after King Julief's experience, Americans learned a similar lesson. When colonial and state governments denied or restricted divorce, serious problems became apparent: mates mistreated each other, were unfaithful, lived separately, abandoned one another, sought annulments for spurious reasons, obtained divorces in more permissive jurisdictions, and subverted the intent of restrained divorce laws by fabricating grounds that fell within legal limitations.

By the first years of the twentieth century, it was apparent that the United States had, like King Julief, accepted divorce as a necessity. Furthermore, the nation had provided a giant petri dish for the growth of divorce. Such factors as industrialization, urbanization, changing roles of women and men, declining commitment to religion, a gradual deterioration of the patriarchal family, rising expectations of marriage, and a long-term belief in personal satisfaction and freedom created a rich medium for the growth of divorce. Thus, in spite of the expenditure of immense amounts of energy and resources to dissuade couples from divorcing, the American divorce rate was higher than those of all other nations.

By the mid-twentieth century, observers inside and outside the United States recognized Americans' propensity to divorce. In 1956, a German commentator labeled the United States "the leading divorce country of the Western world." And in 1971, an American historian prophesied that "divorce in the land of the free will continue to be as American as apple pie."[2]

Yet, even as divorce became prevalent and increasingly acceptable, many people fought its proliferation. Some opponents of divorce maintained marriage was a sacrament and a lifetime agreement, terminable only for the cause stated in the Bible: adultery. Others believed divorce would bring a host of social evils to American society, especially a decline in the institution of the family. A commentator of the early 1980s equated the high divorce rate with what he thought to be a "moral and social crisis" in the United States. In his view, disaster was imminent because "prevailing cultural winds mitigate against marriage."[3]

Some sociologists and other researchers shared the widespread fear that divorce would undermine marriage. They were also alarmed by what appeared to be a waning interest in marriage in the United States. After analyzing the high divorce rate, the rise in numbers of unmarried mothers, and the growing number of never-married Americans, they pronounced the further decline of the American family. Some argued that no-fault legislation had trivialized marriage by making divorce easy. Others speculated that young people of the 1970s and 1980s lacked faith in the continuation of the family and restricted the number of children they bore because they feared eventual divorce.[4]

Despite such assertions, the view of divorce as a necessary, positive force continued to gain support. Increasingly, powerful and articulate forces argued marriage was a dissolvable contract rather than a lifetime agreement. For instance, beginning with the 1971 case of *Boddie v. Connecticut*, Unites States Supreme Court justices ruled that divorce was a citizen's fundamental right. During the 1970s, a number of decisions underwrote freedom of marital choice—and divorce—in a country that believed in individual autonomy.[5]

Shortly after the *Boddie* decision, liberal historian Christopher Lasch stated that divorce actually protected the American family. In 1973, Lasch asserted divorce saved the institution of the family by forcing it to adapt to contemporary conditions and beliefs. Lasch explained that because the traditional patriarchal family was increasingly dysfunctional in twentieth-century America, it was in danger of atrophying altogether. People who divorced traditional mates and sought spouses who agreed equality should characterize marriage created families more suited to the realities of late twentieth-century society. In his view, divorce thus gradually reshaped the family into a more modern, functional form.[6]

By the 1980s, even the Roman Catholic church began to soften its historic anti-divorce stance. Most dioceses and parishes began to provide special programs and weekend retreats for the growing number of divorced Catholics. Also, after a civil divorce either party can request an annulment from the church. An annulment recognizes that a sacramental union was impossible at time of marriage and frees the parties to marry others.[7]

The long-term struggle between those who oppose divorce and those who support it impaired the development of consistent, beneficent divorce laws and policies. Legislators sometimes had to accommodate the two positions in the statutes they enacted, while attorneys, judges, and divorce-seekers often had to manipulate and twist divorce laws to fit individual cases. In addition, because the divorce debate focused on the morality of divorce, such issues as alimony and child custody were frequently shunted aside.

It is, however, possible to reconcile the two views: (1) marriage as a lifetime proposition and (2) divorce as a necessary, and even positive, institution. A life-long marriage can be held as a societal ideal, a goal, a standard to be worked toward, while divorce provisions can end disintegrating marriages incapable of achieving the ideal.

Over forty years ago, anthropologist Margaret Mead informed us that "no matter how free divorce" or "how frequently marriages break up," most societies have assumed marriages would be permanent. Across the world, people have believed that "marriage should last as long as both live." She added that "no known society has ever invented a form of marriage strong enough to stick that did not contain the 'till-death-us-do-part' assumption." Despite this belief, these societies also recognized that some marriages are incapable of lasting a lifetime and provided a mechanism for their dissolution.[8]

If we accept the lesson of the historical record—marriage and divorce exist in virtually all societies—we can also accept the idea that divorce is here to stay. We need not encourage divorce, but we can accept and aid divorce when it occurs.

Because history suggests divorce is unlikely to abate in the United States, it seems sensible to make divorce a constructive institution. If a couple is unable to achieve the goal of marriage as a lifetime endeavor, must one or both individuals suffer as a result of inadequate divorce provisions and lack of services? Or, would it be more sensible to make the divorce process a positive, helpful one?

The past is gone, but it has left in its wake a huge legacy of information and experience—tools that we can use to create what one historian has called "an ethically responsible transition from present to future."[9] In particular, the history of American divorce points to several areas that demand increased attention from legislators, jurists, policy-makers, and individual Americans.

**1.** *Educators and counselors can support the ideal of lifetime marriage and help people avoid divorce by expanding programs designed to educate people before and during marriage.* Of course, educating people about marriage presupposes the existence of a reasonably widespread consensus of what marriage entails. But, as the end of the twentieth century nears, some Americans cling to traditional views of marriage, while others are beginning to embrace new marital models and redefine marriage accordingly.

In 1986, a *McCall's* article described one contemporary model, a marital state called the "new togetherness." It "takes as its criterion the success of the separate life," supports the "independence of husband and wife," and enables "both partners to function well and happily outside the home." Such a marriage would offer different rewards than the traditional breadwinner-homemaker marriage, for a "new togetherness" couple could relate as equals, partners, and friends. They would, as another proponent of this type of marriage phrased it, combine "enduring love with self-development."[10]

Whether they decide to adopt this or other ideals, most Americans would benefit greatly from being forced to examine and adjust their expectations of marriage and to weigh the pros and cons of contemporary models of marriage. Knowing what one expects, and what one wants, in both a mate and marriage is one way of reducing the possibility of divorce.

**2.** *Private and public agencies must help reduce the stresses on contemporary marriages.* Education, counseling, marriage enrichment workshops are already in place and growing, but married couples need more help. For instance, two-job couples with children need reasonable, quality child care. In 1990, an article in *Newsweek* called the state of day care in the United States "haphazard." Its authors pointed out that no federal regulations exist to cover day care: "the government offers consumers more guidance choosing breakfast cereal than child care."[11]

Many policy-makers fear that by supporting child care, federal and state governments would seem to be encouraging women to bear fewer children and to put them increasingly in the hands of others while they work outside their homes. They hesitate to send what seems to be an anti-family message to Americans. In fact, however, the provision and regulation of reliable child care facilities with properly paid employees would proclaim a pro-family stance. Because most American women will hold jobs, either by choice or necessity, some of the services they provided when at home, especially child care, must be provided by others. Government provision and regulation of childcare would assist two-job marriages involving children to remain stable and in force.

**3.** *The mechanism of divorce needs attention and refinement, or perhaps replacement.*   Because both adversary divorce and no-fault divorce have proven problematic, it may be time to devise an entirely new form of divorce. It might be helpful to separate the granting of a divorce decree or dissolution from decisions concerning child custody, alimony, and division of property. Let one court grant the divorce and another determine custody and financial arrangements. As long as custody of children and spousal support are tied to the divorce process, this encourages divorcing spouses to use their children as weapons in their own battles. Clearly, using children in this way increases their bitterness about their parents' divorce and biases them against their own future prospects of achieving happy, lifelong marriages.[12]

**4.** *Alimony, property, and child-care awards must be equitable to all parties involved and must be effectively enforced.*   Intended or not, divorce punishes divorcing men and women. Because husbands and wives fail to achieve the ideal of lifetime marriage, their suffering somehow seems justified. But divorce has especially punished, and continues to, women, many of whom are still economically dependent. Because women, the long-time protectors of home and family, fail to hold crumbling marriages together, their misfortunes seem particularly appropriate.

From colonial times to the present, adequate support for divorcing women and their children has remained a puzzling and unresolved issue. Today, preventing divorced women and their children from entering the ranks of the destitute demands developing an equitable method of dividing a couple's assets, while recognizing the contributions of each partner, even if only one is employed outside the home or one earns less than the other.[13] This requires the enactment of statutes to guide courts in making such decisions and in determining what constitutes assets and contributions. A man's work experience is an asset; a woman's homemaking skills a contribution. Both must be taken into account and valued at the time of divorce.

In addition, strict enforcement procedures must be put in place to ensure that alimony and support payments are made, perhaps through automatic payroll deductions. Special courts and stringent enforcement procedures are the costs that a society based on inequities between men and women must pay. When women receive the same encouragement, education, training, and wages as men, and when child support costs are fairly assessed and effectively enforced, the financial condition of divorcing women will cease to be such a serious predicament in the United States.

Division of Social Security benefits and pensions in divorce also needs more regulation. Despite the pioneering work of Congresswoman Pat Schroeder and the innovative Foreign Service Act of 1980 that provided the first model for the division of pension benefits in case of divorce, many decisions regarding retirement benefits rest in the hands of state courts which are permitted, but not required, to divide such assets between divorcing partners. As a result, a wife, although she has a legal right to part of her husband's military pension, may not receive it. In addition, even if a wife is granted a portion of her husband's retirement benefits, she often discovers that it is virtually impossible to collect.[14] Uniform, strict standards of dividing this form of marital property and an effective payment method are both needed.

In some cases, divorcing men may also need to make claims on their wives' financial resources and on joint property, including pensions and other retirement benefits, especially if they have fewer marketable skills than their wives and/or receive custody of the couple's children. Women's financial post-divorce support to former spouses, and especially to children, requires equitable guiding principles and an enforcement mechanism.

**5.** *Divorced Americans need expanded counseling and other services funded by private organizations and agencies, local and state governments, and the federal government.* Once divorced, Americans continue to require care and support. Mediation and counseling programs, educational and vocational courses, and job placement bureaus for divorcing women and men are already widespread. They encourage divorcing individuals to reject an adversary mentality, to take the best actions for both spouses and their children, to allow themselves some transition and recovery time before establishing new romantic relationships, to determine what they can learn from divorce, and to deal effectively with their children and financial situations.

These programs need expansion as well as increased funding for a number of reasons. In particular, current programs cannot serve everyone who needs help. Also, the current failure rate of remarriages is appallingly high. Yet one of the rationales for divorce is that it allows men and women to contract new marriages that might succeed. If the success rate of remarriages is to rise and a higher number of blended families to remain intact, divorced people must be consoled, counseled, and reconstructed before entering a new marriage. Most are unable to console, counsel, and reconstruct themselves; they must have outside help.

To be effective, counseling and other post-divorce services must develop several important components. A concerted outreach would inform low-income, non-English-speaking, illiterate, and other people that such resources exist. Tailoring services to meet the needs of particular groups, including African Americans, Mexican Americans, and Asian Americans, would make them more effective.[15] And providing free or inexpensive child care would allow both female and male custodial parents to take advantage of these resources without detriment to their children.

Counseling and other services must also develop an awareness of gender differences in clients. Men and women need different advice and forms of assistance. Because women have been urged to base their identity upon their roles as wives, they usually face a greater dislocation than men after divorce. In addition, many women are faced with learning how to be a single parent and with entering, or re-entering, the job market. Often, they must develop a new sense of autonomy and competence. Specialized counseling is necessary if they are to master these transitions successfully.[16]

**6.** *Children of divorce need, and deserve, more study and more counseling services.* Currently, the extent and nature of damage sustained by children of divorce are unclear. Although some experts continue to maintain that children are capable of adjusting to the demands of divorce, others argue that they suffer long-term harm. In a 1988-89 special issue, the *Journal of Divorce* devoted itself to children of divorce. Researchers examined topics ranging from the impact of divorce on children at different stages of the family life cycle to the effects of children's family types on their teachers' images and treatment of them. Also in 1989, Judith Wallerstein and Susan Blakeslee published a ten-year longitudinal study of children of divorce titled *Second Chances* that revealed an appalling amount of long-term damage to children of divorce.[17] Far more studies must be funded and undertaken if we are to understand, and avoid or repair, the harm done to children of divorce, and to reshape the divorce process so it is less traumatic to children.

In addition, counseling services for children of divorce *and* people who work with them must be expanded.[18] For example, a National Fire Prevention Week flyer that is put into children's hands stating that, "when father is home, he is the family fire chief," or an assignment to describe the daily work done by one's father, demeans and distresses children of divorce who live with their mothers, especially if they seldom see their fathers. Because such unintentional cruelty increases the suffering of children of divorce, it must be avoided.

Granted, many changes have taken place since a Puritan magistrate parted the Luxfords in 1639. As divorce grew and became a traditional way of dealing with marital distress, Americans slowly became aware of its perils. During the late twentieth century, they proposed and implemented such reforms as no-fault divorce. They have also tried to remedy such problems as awarding of alimony, division of property, custody of children, child support, and post-divorce education and services.

But more changes need to occur. Fears about the demise of the American family need not hold us back; the family has remained vital in the face of rising divorce rates. Rather than declining as has so often been predicted, the family survives even the most liberal divorce policies. A case in point is Eskimo societies of North Alaska which allow free divorce. There, people continue to marry, while divorced spouses and children form new family units.[19]

In other parts of the United States, where divorce has become easy, marriage and the family also continue to thrive. Apparently, undeterred by the rising divorce rate, Americans marry and remarry. Among the divorced men and women interviewed for this study, 87 percent said that they would remarry if they had the opportunity.[20]

Clearly, Americans have maintained their faith in marriage and the family. Ninety-one percent of interviewees believed that both were viable institutions in the contemporary United States. Of course, part of the vitality of these institutions comes from their ability to expand and to encompass the idealized two-parent, two-child model as well as blended, single-parent, complex, gay and lesbian, and "skip-generation" families.[21]

History suggests that the family will continue to survive and adapt, but it also indicates that divorce will continue to spread. *Americans will divorce.* Thus, American divorce must be made a more constructive, productive institution for the millions of wives, husbands, children, relatives, friends, employers, and others who are touched, and too often scarred, by it.

# Bibliography

**Books**

Bailey, Beth L. *From Front Porch to Backseat: Courtship in Twentieth-Century America.* Baltimore: Johns Hopkins University Press, 1988.

Barnett, James Harwood. *Divorce and the American Divorce Novel, 1858–1937.* New York: Russell & Russell, 1968 ed.

Basch, Norma. *In the Eyes of the Law: Women, Marriage and Property in Nineteenth-Century New York.* Ithaca: Cornell University Press, 1982.

Bernard, Jessie. *The Future of Marriage.* New York: Bantam Books, 1973.

Blake, Nelson Manfred. *The Road to Reno: A History of Divorce in the United States.* New York: Macmillan, 1962.

Bohannan, Paul, ed. *Divorce and After.* Garden City: Doubleday & Co., 1970.

Cahen, Alfred. *Statistical Analysis of American Divorce.* New York: Columbia University Press, 1932.

Cancian, Francesca. *Love in America: Gender and Self-Development.* Cambridge: Cambridge University Press, 1987.

Carter, Hugh, and Paul C. Glick. *Marriage and Divorce: A Social and Economic Study.* Cambridge: Harvard University Press, 1970.

Davis, Kingsley, and Amyra Grossbard-Schechtman. *Contemporary Marriage: Comparative Perspectives on a Changing Institution.* New York: Russell Sage Foundation, 1985.

Degler, Carl N. *At Odds: Women and the Family in America from the Revolution to the Present.* New York: Oxford University Press, 1980.

D'Emilio, John, and Estelle B. Freedman. *Intimate Matters: A History of Sexuality in America.* New York: Harper & Row, 1988.

Demos, John. *A Little Commonwealth: Family Life in Plymouth Colony.* New York: Oxford University Press, 1970.

Despert, Louise J. *Children of Divorce.* New York: Doubleday Dolphin Books, 1953.

Ditzion, Sidney. *Marriage, Morals, and Sex in America: A History of Ideas.* New York: W. W. Norton & Co., 1978 edition.

Ehrenreich, Barbara, and Deidre English. *For Her Own Good: 150 Years of Experts' Advice to Women.* New York: Anchor Books, 1978.

Foster, Henry H., Jr., and Doris Jonas Freed. *The Divorce Reform Law.* New York: The Lawyers Co-operative Publishing Company, 1970.

Gordon, Linda. *Heroes of Their Own Lives: The Politics and History of Family Violence, Boston, 1880–1960.* New York: Viking Press, 1988.

Gordon, Michael, ed., *The American Family in Social-Historical Perspective.* New York: St. Martin's Press, 1973.

Greene, Everts B., and Virginia D. Harrington. *American Population Before the Census of 1790.* Gloucester, Mass.: Peter Smith, 1966.

Griswold, Robert L. *Family and Divorce in California, 1850–1890: Victorian Illusions and Everyday Realities.* Albany: State University of New York Press, 1982.

Grossberg, Michael. *Governing the Hearth: Law and the Family in Nineteenth-Century America.* Chapel Hill: University of North Carolina Press, 1985.

Halem, Lynne Carol. *Divorce Reform: Changing Legal and Social Perspectives.* New York: Free Press, 1980.

Horstman, Allen. *Victorian Divorce* (concerning divorce in England). New York: St. Martin's Press, 1985.

Howard, George E. *A History of Matrimonial Institutions,* 3 vols. Chicago: University of Chicago Press, 1904.

Jacob, Herbert. *Silent Revolution: The Transformation of Divorce Law in the United States.* Chicago: University of Chicago Press, 1988.

Jacobson, Paul H. *American Marriage and Divorce.* New York: Rinehart & Co., 1969.

Johnson, Julia, ed. *Selected Articles on Marriage and Divorce.* New York: H. W. Wilson, 1925.

Kern, Louis J. *An Ordered Love: Sex Roles and Sexuality in Victorian Utopias—the Shakers, the Mormons, and the Oneida Community.* Chapel Hill: University of North Carolina Press, 1981.

Koehler, Lyle. *A Search for Power: The "Weaker Sex" in Seventeenth-Century New England.* Urbana: University of Illinois Press, 1980.

Lantz, Herman R. *Marital Incompatibility and Social Change in Early America.* Beverly Hills: Sage Publications, 1976.

Leach, William. *True Love and Perfect Union: The Feminist Reform of Sex and Society.* New York: Basic Books, 1980.

Lystra, Karen. *Searching the Heart: Women, Men, and Romantic Love in Nineteenth-Century America.* New York: Oxford University Press, 1989.

Luck, William F. *Divorce and Remarriage: Recovering the Biblical View.* New York: Harper and Row, 1987.

May, Elaine Tyler. *Great Expectations: Marriage and Divorce in Post-Victorian America.* Chicago: University of Chicago Press, 1980.

May, Elaine Tyler. *Homeward Bound: American Families in the Cold War Era.* New York: Basic Books, 1988.

Mead, Margaret. *Male and Female: A Study of the Sexes in a Changing World.* New York: William Morrow & Company, 1949.

Mintz, Steven. *A Prison of Expectations: The Family in Victorian Culture.* New York: New York University Press, 1983.

Mintz, Steven and Susan Kellogg. *Domestic Revolutions: A Social History of American Family Life.* New York: Free Press, 1988.

Muncy, Raymond Lee. *Sex and Marriage in Utopian Communities: 19th-Century America.* Bloomington: Indiana University Press, 1973.

O'Neill, William L. *Divorce in the Progressive Era.* New Haven: Yale University Press, 1967.

Perkin, Joan. *Women and Marriage in Nineteenth-Century England.* Chicago: Lyceum Books, 1989.

Petrik, Paula. *No Step Backward: Women and Family on the Rocky Mountain Mining Frontier, Helena, Montana, 1865–1900.* Helena: Montana Historical Society Press, 1987 (Chapter 4).

Phillips, Roderick. *Putting Asunder: A History of Divorce in Western Society.* Cambridge: University of Cambridge Press, 1988.

Pleck, Elizabeth H. *Domestic Tyranny: The Making of Social Policy Against Family Violence from Colonial Times to the Present.* New York: Oxford University Press, 1987.

Rheinstein, Max. *Marriage Stability, Divorce, and the Law.* Chicago: University of Chicago Press, 1972.

Rothman, David J., and Sheila M. Rothman, eds. *Divorce: The First Debates.* New York: Garland Publishing, 1987.

Rothman, Ellen K. *Hands and Hearts: A History of Courtship in America.* New York: Basic Books, 1984.

Salmon, Marylynn. *Women and the Law of Property in Early America.* Chapel Hill: University of North Carolina Press, 1986.

Scott, Donald, and Bernard Wishy, eds. *America's Families: A Documentary History.* New York: Harper and Row, 1981.

Swerdlow, Amy, Renate Bridenthal, Joan Kelly, and Phyllis Vine. *Household and Kin: Families in Flux.* Old Westbury, N.Y.: The Feminist Press, 1981.

Weisberg, D. Kelley. *Women and the Law: The Social Historical Perspective,* 2 vols. Cambridge, Mass.: Schenkman Publishing Co., 1982.

Weitzman, Lenore J. *The Divorce Revolution: The Unexpected Social and Economic Consequences for Women and Children in America.* New York: Free Press, 1985.

Westermarck, Edward. *The History of Human Marriage.* London: Macmillan, 1921.

Wires, Richard. *The Divorce Issue and Reform in Nineteenth-Century Indiana.* Muncie, Ind.,: Ball State University Press, 1967.

**Articles**

Aaron, Richard I., "Mormon Divorce and the Statute of 1852: Questions for Divorce in the 1980s," *Journal of Contemporary Law* 8 (1982):5–45.

Alexander, Jay W., "Conflict of Laws," *North Carolina Law Review* 27, 1 (Dec. 1948):134–41 (implications of *Haddock v. Haddock*, 1906–48).

Arendell, Terry J., "Women and the Economics of Divorce in the Contemporary United States," *Signs* 13, 1 (Autumn 1987):121–35.

Bailey, Beth L., "Scientific Truth . . . and Love: The Marriage Education Movement in the United States," *Journal of Social History* 20, 4 (Summer 1987):711–32.

Bandel, Betty, "What the Good Laws of Man Hath Put Asunder," *Vermont History* 46, 4 (1978):221–33.

Basch, Françoise, "Women's Rights and the Wrongs of Marriage in Mid-Nineteenth Century America," *History Workshop Journal* 22, 3 (Autumn 1986):18–40.

Basch, Norma, "Invisible Women: The Legal Fiction of Marital Unity in Nineteenth-Century America," *Feminist Studies* 5, 2 (Summer 1979):346–66.

Basch, Norma, "The Emerging Legal History of Women in the United States: Property, Divorce, and the Constitution," *Signs* 12, 1 (Autumn 1986):97–117.

Basch, Norma, "Relief in the Premises: Divorce as a Woman's Remedy in New York and Indiana, 1815–1870," *Law and History Review,* forthcoming.

Bozeman, Robert M., "The Supreme Court and Migratory Divorce: A Reexamination of an Old Problem," *American Bar Association Journal* 37, 1 (Feb. 1951):107–10, 168–71.

Brevda, William. "Love's Coming-of-Age: The Upton Sinclair–Harry Kemp Divorce Scandal," *North Dakota Quarterly* 51, 2 (1983):59–77.

Campbell, Eugene E., and Bruce L. Campbell. "Divorce Among Mormon Polygamists: Extent and Explanations," *Utah Historical Quarterly* 46, 1 (1978):4–23.

Censer, Jane Turner, " 'Smiling Through Her Tears': Ante-bellum Southern Women and Divorce," *American Journal of Legal History* 25, 1 (Jan. 1981):24–47.

Chavis, William M., and Gladys J. Lyles. "Divorce Among Educated Black Women," *Journal of the National Medical Association* 67, 2 (March 1975):128–34.

Cheng, C. K., and Douglas S. Yamamura. "Interracial Marriage and Divorce in Hawaii," *Social Forces* 36, 1 (Oct. 1957):77–84.

Clark, Elizabeth, "Matrimonial Bonds: Slavery, Contract and the Law of Divorce in Nineteenth-Century America," Legal History Program Working Papers, Series 1, Number 10, June 1987, Institute for Legal Studies, University of Wisconsin–Madison Law School.

Cohn, Henry S., "Connecticut's Divorce Mechanism: 1636–1639," *American Journal of Legal History* 14, 1 (1970): 35–54.

Cohen, Sheldon S., " 'To Parts of the World Unknown': The Circumstances of Divorce in Connecticut, 1750–1797," *Canadian Review of American Studies* 11, 3 (Winter 1980):275–93.

Cohen, Sheldon S. "The Broken Bond: Divorce in Providence County, 1749–1809," *Rhode Island History* 44, 3 (Aug. 1985):67–79.

Cott, Nancy F., "Eighteenth-Century Family and Social Life Revealed in

Massachusetts Divorce Records," *Journal of Social History* 10, 1 (1976):20–43.

Cott, Nancy F., "Divorce and the Changing Status of Women in Eighteenth-Century Massachusetts," *William and Mary Quarterly* 33, 4 (1976):586–614.

Davis, Kingsley, "Statistical Perspectives on Marriage and Divorce," *Annals of the American Academy of Political and Social Science* 272 (Nov. 1950):9–21.

Driedger, Leo, Michael Yoder, and Peter Sawatzky. "Divorce Among Mennonites: Evidence of Family Breakdown," *Mennonite Quarterly Review* 59, 4 (1985):367–82.

Dunfey, Julie, " 'Living the Principle' of Plural Marriage: Mormon Women, Utopia, and Female Sexuality in the Nineteenth Century," *Feminist Studies* 10, 3 (Fall 1984):523–36.

Eberstein, Isaac W., and W. Parker Frisbie. "Differences in Marital Instability among Mexican Americans, Blacks, and Anglos: 1960 and 1970," *Social Problems* 23, 4 (1976):609–21.

Einhorn, Jay, "Child Custody in Historical Perspective: A Study of Social Perceptions of Divorce and Child Custody in Anglo-American Law," *Behavioral Sciences and the Law* 4, 2 (Spring 1986):119–35.

Fenelon, William, "State Variations in United States Divorce Rates," *Journal of Marriage and the Family* 33, 2 (May 1971):321–27.

Foster, Henry H., Jr., "Current Trends in Divorce Law," *Family Law Quarterly* 1, 2 (June 1967):21–40.

Foster, Henry H., Jr., "For Better or Worse? Decisions Since *Haddock v. Haddock*," *American Bar Association Journal* 47, 3 (Oct. 1961):963–67.

Foster, Lawrence, "Free Love and Feminism: John Humphrey Noyes and the Oneida Community," *Journal of the Early Republic* 1, 2 (Summer 1981):165–83.

Franklin, Carl M., "The Dilemma of Migratory Divorces: A Partial Solution Through Federal Legislation," *Oklahoma Law Review* 1, 3 (Aug. 1948):151–70 (covers 1869 to 1948).

Freed, Doris Jonas, and Henry H. Foster, Jr., "Divorce American Style," *Annals of the American Academy of Political and Social Science* 383 (May 1969):71–88.

Freed, Doris Jonas and Henry H. Foster, Jr., "Divorce in the Fifty States: An Outline," *Family Law Quarterly* 11, 3 (Fall 1977):297–313.

Friedman, Lawrence M., "Rights of Passage: Divorce Law in Historical Perspective," *Oregon Law Review* 63, 4 (1984):649–75.

Friedman, Lawrence W., and Robert V. Percival, "Who Sues for Divorce? From Fault through Fiction to Freedom," *Journal of Legal Studies* 5, 1 (Jan. 1976):61–82.

Frierson, J. Nelson, "Divorce in South Carolina," *North Carolina Law Review* 9, 3 (April 1931): 271–72, 275–82.

Frisbie, W. Parker, Frank D. Bean, and Isaac W. Eberstein, "Recent Changes in Marital Instability among Mexican Americans: Convergence with Black and Anglo Trends?" *Social Forces* 58, 4 (1980):1205–20.

Gadlin, Howard, "Private Lives and Public Order: A Critical View of the History of Intimate Relations in the United States," *Massachusetts Review* 17, 2 (Summer 1976):304–30.

Garfield, Robert, "The Decision to Remarry," *Journal of Divorce* 43, 1 (Fall 1980):1–10.

Glenn, Myra C., "Wife-Beating: The Darker Side of Victorian Domesticity," *Canadian Review of American Studies* 15, 1 (Spring 1984):17–33.

Goodheart, Lawrence B., Neil Hanks, and Elizabeth Johnson, " 'An Act for the Relief of Females . . .': Divorce and the Changing Legal Status of Women in Tennessee, 1796–1860," *Tennessee Historical Quarterly,* Part I, 44, 3 (Fall 1985):318–39; and Part II, 44, 4 (Winter 1985):402–16.

Griffith, Elisabeth, "Elizabeth Cady Stanton on Divorce: Feminist Theory and Domestic Experience," 233–52, in Mary Kelley, ed., *Woman's Being, Woman's Place.* Boston: G. K. Hall, 1979.

Griswold, Robert L., "Apart but not Adrift: Wives, Divorce, and Independence in California, 1850–1890," *Pacific Historical Review* 49, 2 (1980):265–83.

Griswold, Robert L., "Adultery and Divorce in Victorian America, 1800–1900," Legal History Program Working Papers, Series 1, Number 6, March, 1986, Institute for Legal Studies, University of Wisconsin–Madison Law School.

Griswold, Robert L., "The Evolution of the Doctrine of Mental Cruelty in Victorian American Divorce, 1790–1900," *Journal of Social History* 20, 1 (Fall 1986):127–48.

Griswold, Robert L., "Law, Sex, Cruelty, and Divorce in Victorian America, 1840–1900," *American Quarterly* 38, 5 (Winter 1986):721–45.

Griswold, Robert L., "Sexual Cruelty and the Case for Divorce in Victorian America," *Signs* 11, 3 (1986):529–41.

Grossberg, Michael, "Who Gets the Child? Custody, Guardianship, and the Rise of a Judicial Patriarchy in Nineteenth-Century America," *Feminist Studies* 9, 2 (Summer 1983):235–60.

Hampton, Robert L., "Husband's Characteristics and Marital Disruption in Black Families," *Sociological Quarterly* 20, 2 (1979):255–66.

Haugland, Lonnie R., "South Dakota Divorce Legislation and Reform, 1862–1908," *The Region Today: A Quarterly Journal of the Social Science Research Associates* 2, 3 (May 1975):48–65.

Hindus, Michael S., and Lynne E. Withey, "The Law of Husband and Wife in Nineteenth-Century America," 133–54, in D. Kelley Weisberg, ed., *Women and the Law: A Social Historical Perspective.* Cambridge, Mass.: Schenkman Publishing Co., 1982.

Hirsch, Alison Duncan, "The Thrall Divorce Case: A Family Crisis in Eighteenth-Century Connecticut," in Linda E. Speth and Alison Duncan Hirsch, eds., *Women, Family, and Community in Colonial America: Two Perspectives.* New York: Haworth Press, 1983.

Jacobson, Paul H., "Differentials in Divorce by Duration of Marriage and Size of Family," *American Sociological Review* 15, 2 (April 1950):235–44.

Jones, Shirley Maxwell, "Divorce and Remarriage: A New Beginning, a New Set of Problems," *Journal of Divorce* 2, 2 (Winter 1978):217–27.

Kern, Louis J., "Ideology and Reality: Sexuality and Women's Status in the Oneida Community," *Radical History Review* 20, 8 (Spring/Summer 1979):180–204.

Krom, Howard A., "California's Divorce Law Reform: An Historical Analysis," *Pacific Law Journal* 1, 3 (Jan. 1970):156–81.

Lantz, Herman R., Margaret Britton, Raymond Schmitt, and Eloise C. Synder, "Pre-Industrial Patterns in the Colonial Family in America: A Content Analysis of Colonial Magazines," *American Sociological Review* 33, 3 (June 1968):413–26.

Lantz, Herman R., Raymond L. Schmitt, and Richard Herman, "The Pre-industrial Family in America: A Further Examination of Early Magazines," *American Journal of Sociology* 79, 3 (Nov. 1973):566–88.

Lantz, Herman R., Jane Keyes, and Martin Schultz, "The Family in the Preindustrial Period: From Baselines in History to Social Changes," *American Sociological Review* 40, 1 (Feb. 1975):21–36.

Lantz, Herman, Martin Schultz, and Mary O'Hara, "The Changing American Family from the Preindustrial to the Industrial Period: A Final Report," *American Sociological Review* 42, 3 (June 1977):406–21.

Lasch, Christopher, "Divorce and the 'Decline of the Family'," in *The World of Nations: Reflections on American History, Politics, and Culture.* New York: Vintage Books, 1973.

Littlefield, Daniel F., Jr., and Lonnie E. Underhill, "Divorce Seekers' Paradise: Oklahoma Territory, 1890–1897," *Arizona and the West* 17, 1 (1975):21–34.

Lojek, Helen, "The Southern Lady Gets a Divorce: 'Saner Feminism' in the Novels of Amelie Rives," *Southern Literary Journal* 12, 1 (1979):47–69.

May, Elaine Tyler, "The Pressure to Provide: Class, Consumerism, and Divorce in Urban America, 1880–1920," *Journal of Social History* 12, 2 (1978):180–93.

May, Elaine Tyler, "In-Laws and Out-Laws: Divorce in New Jersey, 1890–1925," 31–41 in Mary R. Murrin, ed., *Women in New Jersey History.* Trenton: New Jersey Historical Commission, 1985.

McDonald, Brenda D., "Domestic Violence in Colonial Massachusetts," *Historical Journal of Massachusetts* 14, 1 (Jan. 1986):53–64.

Meehan, Thomas R., " 'Not Made Out of Levity': Evolution of Divorce in Early Pennsylvania," *Pennsylvania Magazine of History and Biography* 92, 4 (1968):441–64.

Munson, C. La Rue, "The Divorce Question in the United States," *Yale Law Journal* 18, 6 (April 1909):387–98.

Nadelhaft, Jerome, "Wife Torture: A Known Phenomenon in Nineteenth-Century America," *Journal of American Culture* 10, 3 (Fall 1987):39–59.

Nolan, Val, Jr., "Indiana: Birthplace of Migratory Divorce," *Indiana Law Journal* 26, 3 (Summer 1951):515–27.

Norton, Arthur J., and Paul C. Glick, "Marital Instability: Past, Present and Future," *Journal of Social Issues* 32, 1 (1976):5–20.

O'Neill, William L., "Divorce as a Moral Issue: A Hundred Years of Controversy," 127–43, in Carol V. R. George, ed., *'Remember the Ladies': New Perspectives on Women in American History.* Syracuse: Syracuse University Press, 1975.

O'Neill, William L., "Divorce in the Progressive Era," *American Quarterly* 17, 2 (1965):203–17.

Pang, Henry, and Sue Marie Hanson, "Highest Divorce Rates in Western United States," *Sociology and Social Research* 52, 2 (Jan. 1968):228–36.

Pearson, Willie, Jr., and Lewellyn Hendrix, "Divorce and the Status of Women," *Journal of Marriage and the Family* 41, 2 (May 1979):375–87.

Petrik, Paula, " 'If She Be Content': The Development of Montana Divorce Law, 1865–1907," *Western Historical Quarterly* 23, 3 (July 1987):261–92.

Preston, Samuel H., and John McDonald, "The Incidence of Divorce Within Cohorts of American Marriages Contracted Since the Civil War," *Demography* 16, 1 (Feb. 1979):1–26.

Raphael, Robert, Frederick N. Frank, and Joanne Ross Wilder, "Divorce in America: The Erosion of Fault," *Dickinson Law Review* 81, 3 (Summer 1977):719–31.

Reynolds, Robert R., Jr., et al., "An Exploratory Analysis of County Divorce Rates," *Sociology and Social Research* 69, 1 (1984):109–21.

Rheinstein, Max, "The Law of Divorce and the Problem of Marriage Stability," *Vanderbilt Law Review* 9, 4 (June 1956):633–34.

Riddell, William Renwick, "Legislative Divorce in Colonial Pennsylvania," *Pennsylvania Magazine of History and Biography* 57, 2 (1933):175–80.

Robinson, Doane, "Divorce in Dakota," *South Dakota Historical Collections,* 12, 1924, 268–80. Pierre: Hipple Printing Company, 1924.

Ross, Ellen, " 'The Love Crisis': Couples Advice Books of the Late 1970s," *Signs* 6, 1 (Autumn 1980):109–22.

Ruymann, William G., "The Effects of Nevada Divorce Decrees Out of the State," *Nevada State Bar Journal* 14, 4 (Oct. 1949):239–55.

Sander, William, "Women, Work, and Divorce," *American Economic Review* 75, 3 (1985):519–23.

Sell, Kenneth D., "Divorce Advertising—One Year After *Bates.*" *Family Law Quarterly* 12, 4 (Winter 1979):275–83.

Sepler, Harvey J., "Measuring the Effects of No-Fault Divorce Laws Across Fifty States: Quantifying a Zeitgeist," *Family Law Quarterly* 15, 1 (Spring 1981):65–102.

Shientag, Florence Perlow, "Divorce in the United States: Recent Developments," *Women Lawyers Journal* 43, 1 (Winter 1957):7–9, 20–21.

Smith, Daniel Scott, and Michael S. Hindus, "Premarital Pregnancy in America, 1640–1971: An Overview and Interpretation," *Journal of Interdisciplinary History* 5, 4 (Spring 1975):537–70.

Smith, Daniel Scott, "Parental Power and Marriage Patterns: An Analysis of Historical Trends in Hingham, Massachusetts," *Journal of Marriage and the Family* 35, 3 (Aug. 1973):419–28.

Smith, Daniel Blake, "The Study of the Family in Early America: Trends,

Problems, and Prospects," *William and Mary Quarterly* 39, 1 (Jan. 1982):3–28.

Spalletta, Matteo, "Divorce in Colonial New York," *New-York Historical Society Quarterly* 39, 3 (Oct. 1955):422–40.

Squire, George, "The Shift from Adversary to Administrative Divorce," *Boston University Law Review* 33, 2 (April 1953):141–75.

Stetson, Dorothy M., and Gerald C. Wright, Jr., "The Effects of Laws on Divorce in American States," *Journal of Marriage and the Family* 37, 3 (Aug. 1975):537–47.

Stone, Lawrence, "Family History in the 1980s: Past Achievements and Future Trends," *Journal of Interdisciplinary History* 12, 1 (Summer 1981):51–87.

Strickman, Leonard P., "Marriage, Divorce and the Constitution," *Family Law Quarterly* 14, 4 (Winter 1982):259–348.

Sumner J. D., Jr., "The South Carolina Divorce Act of 1949," *South Carolina Law Quarterly* 3, 3 (March 1951):253–59.

Sweet, James A., and Larry L. Bumpass, "Differentials in Marital Instability of the Black Population, 1970," *Phylon* 35, 3 (Sept. 1974):323–31.

Swisher, Peter Nash, "Foreign Migratory Divorces: A Reappraisal," *Journal of Family Law* 21, 1 (1982–83):9–52.

Thornton, Arland, "Changing Attitudes Toward Separation and Divorce: Causes and Consequences," *American Journal of Sociology* 90, 4 (Jan. 1985):856–72.

Traill, David H., "Schliemann's American Citizenship and Divorce," *Classical Journal* 77, 4 (May/June 1982):336–42.

Uba, George R., "Status and Contract: The Divorce Dispute of the 'Eighties and Howells' *A Modern Instance,*" *Colby Library Quarterly* 19, 2 (June 1983):78–89.

Uhlenberg, Peter, "Marital Instability among Mexican Americans: Following the Pattern of Blacks?" *Social Problems* 20, 1 (1972):49–56.

Van Ness, James S., "On Untieing the Knot: The Maryland Legislature and Divorce Petitions," *Maryland Historical Magazine* 67, 2 (Summer 1972):171–75.

Vliet, R. Dale, "A Foreseeable End to Migratory Divorces," *Oklahoma Law Review* 10, 4 (Nov. 1957):432–38.

Walroth, Joanne Ruth, "Beyond Legal Remedy: Divorce in Seventeenth-Century Woodbridge, New Jersey," *New Jersey History* 105, 3/4 (Fall/Winter 1987):1–36.

Weisbrod, Carol, and Pamela Sheingorn, "*Reynolds v. United States:* Nineteenth-Century Forms of Marriage and the Status of Women," *Connecticult Law Review* 10, 3 (Summer 1978):828–58.

Weisberg, D. Kelly, " 'Under Greet Temptations Heer': Women and Divorce in Puritan Massachusetts," *Feminist Studies* 2, 2/3 (1975):183–93.

Wells, Robert V., "Demographic Change and the Life Cycle of American Families," *Journal of Interdisciplinary History* 2, 2 (Autumn 1971):273–82.

Wolfram, Sybil, "Divorce in England, 1700–1857," *Oxford Journal of Legal Studies* 5, 2 (Summer 1985):155–86.

Young, Rev. James J., "The Divorced Catholics Movement," *Journal of Divorce* 2, 1 (Fall 1978):83–97.

Zenor, Donna J., "Untying the Knot: The Course and Patterns of Divorce Reform," *Cornell Law Review* 57, 2 (April 1972):649–67.


**Theses and Dissertations**

Ford, Bonnie L., "Women, Marriage, and Divorce in California, 1849–1872," Ph.D. dissertation, University of California, Davis, 1985.

Frobish, Dennis Lee, "The Family and Ideology: Cultural Constraints on Women, 1940–1960," Ph.D. dissertation, University of North Carolina, Chapel Hill, 1983.

Hult, Susan Freda, "Divorce and the Decline of Patriarchy in South Carolina," M.A. thesis, Clemson University, Clemson, S.C., 1985.

Jones, Mary Somerville, "An Historical Geography of Changing Divorce Law in the Unites States," Ph.D. dissertation, University of North Carolina, Chapel Hill, 1978.

Labbé, Dolores Egger, "Women in Early Nineteenth-Century Louisiana," Ph.D. dissertation, University of Delaware, 1975.

May, Elaine Tyler, "The Pursuit of Domestic Perfection: Marriage and Divorce in Los Angeles, 1890–1920," Ph.D. dissertation, University of California, Los Angeles, 1975.

Noudeck, Mariellen MacDonald, "Morality Legislation in Early North Dakota, 1889–1914," M.A. thesis, University of North Dakota, Grand Forks, 1964.

Schultz, Martin J., "The Evolution of Marital Disruption in Early America: A Sociological Examination," Ph.D. dissertation, Southern Illinois University, Carbondale, 1980.

Strasser, Myra A., "Social and Cultural Development in 19th-Century Sioux Falls, South Dakota," M.A. Thesis, University of South Dakota, Vermillion, 1969.

# Notes

**Preface**

1. David E. Hunter and Phillip Whitten, eds., *Encyclopedia of Anthropology* (New York, 1976), 217; John Friedl and Michael B. Whiteford, *The Human Portrait*, 2nd ed. (Englewood Cliffs, NJ., 1988), 222; and Michael C. Howard, *Contemporary Cultural Anthropology* 3rd ed. (Glenview, Ill., 1989), 170.


**Introduction**

1. Based on 157 nationwide personal interviews that involved two general questions followed by open-ended discussion conducted by the author between September 19, 1986, and January 16, 1990, with people of varying ages, social classes, educational backgrounds, and occupations, including several historians and other scholars. Ninety-one percent of the respondents had no knowledge of the existence of divorce in colonial America and 57 percent expressed shock that Americans of the 1600s practiced divorce.

2. George E. Howard, *A History of Matrimonial Institutions*, 3 vols (Chicago, 1904); and Nelson Manfred Blake, *The Road to Reno: A History of Divorce in the United States* (New York, 1962).

3. For an overview of changing divorce law see Lawrence M. Friedman, "Rights of Passage: Divorce Law in Historical Perspective," *Oregon Law Review* 63 (1984):649–69; Max Rheinstein, *Marriage Stability, Divorce, and the Law* (Chicago, 1972); and Roderick Phillips, *Putting Asunder: A History of Divorce in Western Society* (Cambridge, Eng., 1988).

4. Two recent and important interpretations of causes of divorce are Elaine Tyler May, *Great Expectations: Marriage and Divorce in Post-Victorian America* (Chicago, 1980) and Robert L. Griswold, *Family and Divorce in California, 1850–1890: Victorian Illusions and Everyday Realities* (Albany, 1982).

5. U.S. Dept. of Commerce and Labor, *Marriage and Divorce, 1867–1906* (Westport, Conn., reprint ed. 1978, original 1909), I, 22.

6. United Nations, *1986 Demographic Yearbook* (New York, 1988), 470–73. See also National Center for Health Statistics, *Vital Statistics of the United States, 1984,* Vol. III, *Marriage and Divorce* (Washington, D. C., 1988), Sect. 2, p. 5; and Hugh Carter and Paul C. Glick, *Marriage and Divorce: A Social and Economic Study* (Cambridge, Mass., 1970), 17.

7. For a discussion of various types of divorce legislation policy-making processes, especially the theory that such laws are often routine measures to make divorce statutes consistent with practice, see Herbert Jacob, *Silent Revolution: The Transformation of Divorce Law in the United States* (Chicago, 1988).

8. For a discussion of the state's role in marriage and divorce see Susan C. Nicholas, Alice M. Price, and Rachel Rubin, *Rights and Wrongs: Women's Struggle for Legal Equality* (Old Westbury, N. Y., 1979), 22–29; and Harold A. Meriam, Jr., "Estoppel in Matrimonial Actions—A Survey of the New York Policy," *New York Law Review* 14 (April 1948):254–60.

9. See Lenore J. Weitzman, *The Divorce Revolution: The Unexpected Social and Economic Consequences for Women and Children in America* (New York, 1985). The pros and cons of no-fault divorce are discussed further in Chapter 7 below.

10. Judith S. Wallerstein and Sandra Blakeslee, *Second Chances: Men, Women, and Children a Decade After Divorce* (New York, 1989).

11. Not only do the majority of Americans marry, but more than three-fourths of divorced Americans remarry, so that currently approximately 30 percent of brides and 32 percent of grooms have been previously divorced. U.S. Dept. of Health and Human Services, *Remarriages and Subsequent Divorces: United States* (Hyattsville, Md., 1989), 1–2, 5.

12. See Ernest W. Burgess, *The Family* (New York, 1945), 630, for a discussion of the role played by national attitudes in determining divorce rates.

### Chapter 1

1. One example of marital upset was the 1661 case of Mary Andrewes who asked to "be freed from her husband" because she had heard that he had been previously married to a woman now living in Ireland. *Records of the Colony or Jurisdiction of New Haven, 1653–1665* (Hartford, 1858), II, 425–27. Newspapers surveyed were the *Boston Evening Post,* 1735–75, and the *South Carolina Gazette,* 1732–55. See also Julia Cherry Spruill, *Women's Life and Work in the Southern Colonies* (Chapel Hill, 1938), 178–84. For a discussion of newspapers and magazines as sources concerning personal relationships see Herman R. Lantz, *Marital Incompatibility and Social Change in Early America* (Beverly Hills, 1976), 5–48.

2. Examples of forced marriages are found in *Plymouth Church Records, 1620–1859,* Part I, Vol. 22, in *Publications of the Colonial Society of Massachu-*

*setts* (Boston, 1920), 197, 216, 251, 256, 261, 269. The New Haven fornication law is in *New-Haven's Settling in New-England and Some Lawes for Government* (London, 1656), 590; the Bedforde case is in *Records of the Colony or Jurisdiction of New Haven, 1653–1665* (Hartford, 1858), I, 89.

3. *Plymouth Church Records, 1620–1859,* Part I, 302–3, 331–32.

4. Ibid., 334; and *Records of the Colony of New Haven,* I, 234. Other cases of marital upheaval are described in Lyle Koehler, *A Search for Power: The "Weaker Sex" in Seventeenth-Century New England* (Urbana, 1980), 152–60.

5. Robert Ashton, ed., *The Works of John Robinson* (Boston, 1851), I, 241. Examples of forced reconciliations are found in *Plymouth Church Records, 1620–1859,* Part I, 258–59, 360–61. Forced reconciliations are discussed in John Demos, *A Little Commonwealth: Family Life in Plymouth Colony* (New York, 1970), 91–93.

6. Wilkins is quoted in Steven Mintz and Susan Kellogg, *Domestic Revolutions: A Social History of American Family Life* (New York, 1988), 11.

7. For a listing of instances justifying divorce published in 1702 by Puritan minister Cotton Mather, see Cotton Mather, *Magnalia Christi Americana* (Connecticut, 1853 edition), 253–54.

8. A discussion of the idea that divorce was a protective device is found in K. Kelly Weisberg, " 'Under Greet Temptations Heer': Women and Divorce in Puritan Massachusetts," *Feminist Studies* 2 (1975):186–87; and Marylynn Salmon, *Women and the Law of Property in Early America* (Chapel Hill, 1986), 7–9, 61–62. For descriptions of the Puritan family see Edmund S. Morgan, *The Puritan Family* (New York, 1966), 29–64; Demos, *A Little Commonwealth,* 82–99; Philip J. Greven, Jr., "Family Structure in Seventeenth-Century Andover, Massachusetts," 20–37, in Michael Gordon, ed., *The American Family* (New York, 1978) and Mintz and Kellogg, *Domestic Revolutions,* 6–13.

9. Examples of the English view of the sanctity and indissolvable nature of marriage are Daniel Defoe, *Conjugal Lewdness; or, Matrimonial Whoredom* (Gainesville, Fla., 1967, original 1727) and Thomas Grantham, *A Wife Mistaken, or, A Wife and No Wife* (London, ca. 1750). The English adhered to the biblical view of marriage and divorce which is discussed in detail in William F. Luck, *Divorce and Remarriage: Recovering the Biblical View* (San Francisco, 1987).

10. An early and strong argument for civil marriage in England was *Reformatio Legum Ecclesiasticarum,* a 1552 report on ecclesiastical laws by Puritan divines and lawyers authorized by Parliament but never put into effect. The influence of the views of Martin Luther and John Calvin are discussed in Max Rheinstein, *Marriage Stability, Divorce, and the Law* (Chicago, 1972), 22–27. An important later work was John Milton, *The Doctrine and Discipline of Divorce* (London, 1643). For a discussion of English view on marriage and divorce see George E. Howard, *A History of Matrimonial Institutions* 3 vols. (Chicago, 1904), II, 77; Chilton L. Powell, "Marriage in Early New England," *New England Quarterly* I (July 1928):323–34; and Roderick Phillips, *Putting Asunder: A History of Divorce in Western Society* (Cambridge, Eng., 1988), 95–133.

11. William T. David, ed., *Bradford's History of Plymouth Plantation, 1606–1646* (New York, 1964), 12–13.

12. Koehler, *A Search for Power,* Appendix I, 456.

13. In 1636, Governor John Winthrop mentioned a request for divorce on the grounds of adultery, but did not indicate that a divorce was granted. See John Winthrop, *Winthrop's Journal "History of New England" 1630–1649,* edited by James Kendall Hosmer, 2 vols. (New York, 1908), I, 287. The Luxford case is found in John A. Noble and Joseph F. Cronin, eds., *Records of the Court of Assistants of the Colony of the Massachusetts Bay, 1630–1692,* 3 vols. (Boston, 1901–28), II, 89. Descriptions of additional early cases are in Nancy F. Cott, "Divorce and the Changing Status of Women in Eighteenth-Century Massachusetts," *William and Mary Quarterly* 33 (1976):586–614.

14. *Records of the Court of Assistants,* II, 138.

15. Barbara Aronstein Black, "The Judicial Power and the General Court in Early Massachusetts (1634–1686)," Ph.D. dissertation, Yale University, 1975, p. 195; William H. Whitmore, ed., *The Colonial Laws of Massachusetts* (Boston, 1889), 142; Howard, *Matrimonial Institutions,* II, 331–32; and Charles Cowley, *Our Divorce Courts: Their Origin and History* (Lowell, 1879), 12, 28–31.

16. Demos, *A Little Commonwealth,* 97; and Phillips, *Putting Asunder,* 136–37.

17. For a discussion of the prevalence of husbandly cruelty in Massachusetts see Brenda D. McDonald, "Domestic Violence in Colonial Massachusetts," *Historical Journal of Massachusetts* 14 (Jan. 1986):53–64. The Halsall case is found in Nathaniel B. Shurtleff, ed., *Records of the Governor and Company of the Massachusetts Bay in New England,* 5 vols. (Boston, 1853–54), I, 272, 380, 401.

18. Howard, *Matrimonial Institutions,* II, 337.

19. *Acts and Laws, Passed by the Great and General Court of the Massachusetts Bay* (Boston, 1692), 33.

20. *Acts and Resolves, Public and Private, of the Province of Massachusetts Bay* (Boston, 1869–1922), I, 171–72; John D. Cushing, ed., *Massachusetts Province Laws, 1692–1699* (Wilmington, Del., 1978), 82; and Mather, *Magnalia Christi Americana,* II, 253–54. See also Morgan, *The Puritan Family,* 35–37.

21. Howard, *Matrimonial Institutions,* II, 346, 348. In addition to courts granting divorces to African Americans, white churches occasionally accepted African American men and women as church members although this was against customary practice. See *Plymouth Church Records, 1620–1850,* Part I, Vol. 22, pp. 295, 297.

22. D. Kelly Weisberg, " 'Under Great Temptations Heer': Women and Divorce in Puritan Massachusetts," *Feminist Studies* 2 (1975):186–87.

23. Howard, *Matrimonial Institutions,* II, 338–39; and Cott, "Divorce and the Changing Status of Women," 610.

24. Bridget Hill, ed., *The First English Feminist: Reflections Upon Marriage and Other Writings by Mary Astell* (New York, 1986), 90; Allen Horstman, *Victorian Divorce* (New York, 1985), 21; and Joan Perkin, *Women and Mar-*

*riage in Nineteenth-Century England* (Chicago, 1989), 110–11. See also Sybil Wolfram, "Divorce in England, 1700–1857," *Oxford Journal of Legal Studies* 5 (Summer 1985):155–86.

25. Cott, "Divorce and the Changing Status of Women," 592–96, 599–604, 609–11.

26. Nancy F. Cott, "Eighteenth-Century Family and Social Life Revealed in Massachusetts Divorce Records," *Journal of Social History* 10 (1976):20–43. That romantic love was already a concern in mid-eighteenth-century America is demonstrated in Herman R. Lantz et al., "Pre-Industrial Patterns in the Colonial Family in America: A Content Analysis of Colonial Magazines," *American Sociological Review* 33 (June 1968):413–26. Other discussions of changes in traditional family forms during the 1700s are found in Daniel Scott Smith, "Parental Power and Marriage Patterns: An Analysis of Historical Trends in Hingham, Massachusetts," 87–100, in Michael Gordon, ed., *The American Family in Social-Historical Perspective* (New York, 1973); Daniel Blake Smith, "The Study of the Family in Early America: Trends, Problems, and Prospects," *William & Mary Quarterly* 39 (Jan. 1982):3–28; and Rudy Ray Seward, *The American Family: A Demographic History* (Beverly Hills, Calif., 1978), 37–68.

27. The Lufkin case is described in Cott, "Eighteenth-Century Social Life," 31–34.

28. Figures for divorces are from Howard, *Matrimonial Institutions,* II, 332–33; and Cott, "Divorce and the Changing Status of Women," 592. Koehler's *A Search for Power,* 12–53, 320–21, 362–63, 453–59, also includes figures on divorce in Massachusetts as well as other New England colonies. Robert V. Wells made the point that the death of one spouse ended the majority of marriages in *Revolution in Americans' Lives: A Demographic Perspective on the History of Americans, Their Families, and Their Society* (Westport, Conn., 1982), 42.

29. Figures on divorce are from Howard, *Matrimonial Institutions,* II, 332–33; Cott, "Divorce and the Changing Status of Women," 587, 592–93, 596–604, 613–14; and Phillips, *Putting Asunder,* 249–50.

30. Population statistics are found in Evarts B. Greene and Virginia D. Harrington, *American Population Before the Federal Census of 1790* (Gloucester, Mass., 1966), 13–17, 22.

31. U.S. Dept. of Commerce and Labor, *Marriage and Divorce, 1867–1906* (Westport, Conn., 1978 reprint ed.), I, 22.

32. Drury case is found in the *Records of the Suffolk County Court, 1671–1680* (Boston, 1933), II, 754 and 837.

33. Quoted in Henry S. Cohn, "Connecticut's Divorce Mechanism: 1636–1969," *American Journal of Legal History* 14 (1970):37.

34. Ibid., 37–38; and *New-Haven's Settling . . . and Some Lawes,* 586.

35. *Records of the New Haven Colony,* II, 201–2.

36. Ibid., 209–12.

37. Ibid., 38–39.

38. Salmon, *Women and the Law of Property,* 69.

39. Sheldon S. Cohen, " 'To Parts of the World Unknown': The Circumstances of Divorce in Connecticut, 1750–1797," *The Canadian Review of*

*American Studies* 11 (Winter 1980):277–78; and Cohn, "Connecticut's Divorce Mechanism," 40. For a discussion of the doctrine of mutual fault see Arthur Sherman, "The Doctrine of Recrimination in Massachusetts (Recrimination Rejected)," *Boston University Law Review* 33 (Nov. 1953):454–72.

40. Ibid., 40–41.

41. Alison Duncan Hirsch, "The Thrall Divorce Case: A Family Crisis in Eighteenth-Century Connecticut," *Women, Family, and Community in Colonial America: Two Perspectives* (New York, 1983), 43–75.

42. Cohen, " 'To Parts of the World Unknown,' " 280–81.

43. Ibid.

44. Linda Kerber, *Women of the Republic: Intellect and Ideology in Revolutionary America* (Chapel Hill, 1980), 172; and Cohen, " 'To Parts of the World Unknown,' " 285–86.

45. Cohen, " 'To Parts of the World Unknown,' " 286.

46. Ibid., 275. Although many scholars believe that her diary was written by someone other than Knight, it contains observations that were supposedly hers. See Sarah K. Knight in *The Private Journal of a Journey from Boston to New York in the Year 1704* (Albany, 1865), 55.

47. Cohen, " 'To Parts of the World Unknown,' " 276, 281–82. Estimated black population is found in Greene and Harrington, *American Population*, 50.

48. Howard, *Matrimonial Institutions*, II, 360–66; and Sheldon S. Cohen, "The Broken Bond: Divorce in Providence County, 1749–1809," *Rhode Island History* 44 (Aug. 1985):68.

49. Cohen, "The Broken Bond," 69–76.

50. Quoted, ibid., 73–74.

51. Ibid., 71.

52. Howard, *Matrimonial Institutions*, II, 348–49.

53. Matteo Spalletta, "Divorce in Colonial New York," *New-York Historical Society Quarterly* 39 (Oct. 1955):422.

54. Phillips, *Putting Asunder*, 141.

55. Spalletta, "Divorce in Colonial New York," 428–34.

56. William Nelson, ed., *Documents Relating to the Colonial History of the State of New Jersey*, XXII: *Marriage Records, 1665–1800* (Paterson, 1900), lxxix-lxxxi, lxxxvii-lxxxviii.

57. Joanne Ruth Walroth, "Beyond Legal Remedy: Divorce in Seventeenth-Century Woodbridge, New Jersey," *New Jersey History* 105 (Fall/Winter 1987):1–36.

58. *The Statutes at Large of Pennsylvania from 1682–1801* (Harrisburg, 1896–1908), II, 8–9, 178–84 and VII, 263–65, 606–7, 618; and Thomas R. Meehan, " 'Not Made Out of Levity': Evolution of Divorce in Early Pennsylvania," *Pennsylvania Magazine of History and Biography* 92 (1968):443–46.

59. Samuel Hazard, *Annals of Pennsylvania from the Discovery of the Delaware* (Port Washington, N. Y., 1970), 310; and John D. Cushing, ed., *The Earliest Printed Laws of Delaware, 1704–1741* (Wilmington, 1978), 69–73.

60. Howard, *Matrimonial Institutions*, II, 366–76; and Salmon, *Women and the Law of Property*, 62–63.

61. Spruill, *Women's Life and Work*, 342–44.

62. James S. Van Ness, "On Untieing the Knot: The Maryland Legislature and Divorce Petitions," *Maryland Historical Magazine* 67 (Summer 1972):171–73.

63. Salmon, *Women and the Law of Property*, 62.

64. Ibid., 63.

65. Quoted in Alan D. Watson, "Women in Colonial North Carolina: Overlooked and Underestimated," *North Carolina Historical Review* 68 (Jan. 1981):7.

66. William Saunders and Walter Clark, eds., *Colonial and State Records of North Carolina* (Goldsboro, N. C., 1886–1907), VIII, 228.

67. *Virginia Gazette*, April 7, 1775. See also Spruill, *Women's Life and Work*, 167–84.

68. Marylynn Salmon, "Women and Property in South Carolina: The Evidence from Marriage Settlements, 1730 to 1830," *William and Mary Quarterly* 39 (Oct. 1982):655–85. The Haynes case is described in Watson, "Women in Colonial North Carolina," 9–11.

69. *Louise Jousset La Loire v. Surgeon Pierre de Manade*, Feb. 15, 1728; and *Elizabeth de Villiers v. Francisco de Volsay*, Jan. 27, 1944, Records of the Superior Council of Louisiana, Louisiana Historical Center, New Orleans.

70. Carl J. Eckberg, *Colonial Ste. Genevieve: An Adventure on the Mississippi Frontier* (Gerald, Mo., 1985), 192–95.

71. *Françoise Pery v. Gerard Pery*, Sept. 21, 22, 24, 27, 28, Oct. 26, Dec. 10, 1743, Jan. 27, 1744, Records of the Superior Council of Louisiana, Louisiana Historical Center, New Orleans.

72. Membrede's answer to wife's suit, Sept. 2, 1745, Superior Council of Louisiana, Louisiana Historical Center, New Orleans.

73. Cott, "Divorce and the Changing Status of Women," 590–91.

74. Thomas R. Meehan, " 'Not Made Out of Levity': Evolution of Divorce in Early Pennsylvania," *Pennsylvania Magazine of History and Biography* 92 (1968):444–45.

75. William Renwick Riddell, "Legislative Divorce in Colonial Pennsylvania," *Pennsylvania Magazine of History and Biography* 57 (1933):175–80.

76. Frank L. Dewey, "Thomas Jefferson's Notes on Divorce," *William and Mary Quarterly* 39 (1982):216–19. For middle- and upper-class Virginians' pre-Revolutionary expectations of marriage see Jan Lewis, "Domestic Tranquility and the Management of Emotion among the Gentry of Pre-Revolutionary Virginia," *William and Mary Quarterly* 39 (Jan. 1982):135–49.

77. Cott, "Divorce and the Changing Status of Women," 593; and Cohen, " 'To Parts of the World Unknown,' " 286.

78. Horstman, *Victorian Divorce*, 8.

79. Quoted in Cohen, " 'To Parts of the World Unknown,' " 289; and Kerber, *Women of the Republic*, 180. That a similar rise in premarital pregnancies during the war years also constituted a revolt against patriarchal power is convincingly argued in Daniel Scott Smith and Michael S. Hindus, "Premarital Pregnancy in America, 1640–1971," *Journal of Interdisciplinary History* V (Spring 1975):557–60. The rhetoric of the Revolution is discussed in Edwin G. Burrows and Michael Wallace, "The American Revo-

lution: The Ideology and Psychology of National Liberation," in Donald Fleming and Bernard Bailyn, eds., *Perspectives in American History* (Cambridge, Mass., 1972), VI, 167–267.

80. *The Boston Evening Post*, June 10, 1765, and September 9, 1765.

## Chapter 2

1. Michel Chevalier, *Society, Manners, and Politics in the United States* (Boston, 1839), 415, 429. It should be noted, however, that common law marriages also received far more support in the United States than in England.

2. Quoted in George E. Howard, *A History of Matrimonial Institutions*, 3 vols. (Chicago, 1904), II, 32–35.

3. James S. Van Ness, "On Untieing the Knot: The Maryland Legislature and Divorce Petitions," *Maryland Historical Magazine* 67 (Summer 1972):173–75.

4. Guion G. Johnson, *Ante-Bellum North Carolina: A Social History* (Chapel Hill, 1937), 221–22.

5. W. Magruder Drake, ed., "A Discourse on Divorce: Orleans Territorial Legislature, 1806," *Louisiana History* 22 (1981):437.

6. Jane Turner Censer, " 'Smiling Through Her Tears': Ante-Bellum Southern Women and Divorce," *American Journal of Legal History* 25 (Jan. 1981):note 8, 26.

7. Chapter 64, *Acts Passed at a General Assembly of the Commonwealth of Virginia, 1802* (no place, no date), 46–47.

8. Chapter 6, *Acts Passed at a General Assembly of the Commonwealth of Virginia, 1803* (no place, no date), 20–21.

9. Chapter 59, *Acts Passed at a General Assembly of the Commonwealth of Virginia, 1806* (no place, no date), 26, declared the mulatto child illegitimate, while Chapter 120, *Acts Passed at a General Assembly of the Commonwealth of Virginia, 1816* (Richmond, 1817), 176, did not. The 1814 Jones case is found in Chapter 48, *Acts Passed at a General Assembly of the Commonwealth of Virginia, 1815* (Richmond, 1815), 145.

10. Quoted in James Hugo Johnston, *Race Relations in Virginia & Miscegenation in the South, 1776–1860* (Amherst, 1970), 252–53, 255.

11. The first two cases are described in Johnston, *Race Relations in Virginia* 238–49. The last case is found in Jacqueline Jones, " 'My Mother Was Much of a Woman': Black Women, Work, and the Family Under Slavery," *Feminist Studies* 8 (Summer 1982):248.

12. Chapter 220, *Acts Passed at a General Assembly of the Commonwealth of Virginia, 1817* Richmond, 1818), 220. The prohibition of Mary from remarriage might be interpreted as a warning to wives to maintain good relations with their husbands if at all possible.

13. Chapter 132, *Acts Passed at a General Assembly of the Commonwealth of Virginia, 1819* (Richmond, 1820), 100; and Chapter 222, *Acts of the General Assembly of Virginia, 1841–42* (Richmond, 1842), 163.

14. John W. Williams, comp., *Index of the Enrolled Bills of the General Assembly of Virginia, 1776–1910* (Richmond, 1911), Virginia State Library, Richmond.

15. See, for example, Chapters 42, 43, 49, 50, and 91, *Acts Passed at a General Assembly of the Commonwealth of Virginia, 1808* (no place, no date), 43–44, 50, 57, 85–86.

16. Chapter 223, *Acts of the General Assembly of Virginia, 1841–42.* (Richmond, 1842), 163. Another example is found in Chapter 278, *Acts of the General Assembly of Virginia, 1846–47* (Richmond, 1847), 232.

17. See, for example, Chapter 86, *Acts Passed at a General Assembly of the Commonwealth of Virginia, 1813* (Richmond, 1814), 143 and Chapter 161, *Acts Passed at a General Assembly of the Commonwealth of Virginia, 1818* (Richmond, 1819), 186. The Pettus and Brady cases are found in Chapters 131 and 132, *Acts, 1819,* 99–100.

18. For examples of 1840s cases in which nothing was said about remarriage of either spouse see Chapters 279, 280, 282, *Acts, 1846–47,* 232–33.

19. Chapter 119, *Acts Passed at a General Assembly of the Commonwealth of Virginia, 1816* (Richmond, 1817), 175–76.

20. Chapters 221 and 222, *Acts, 1817,* 220–21. The Evans case is found in Chapter 263, *Acts of the General Assembly of Virginia, 1839* (Richmond, 1839), 199. For 1840s cases in which mothers got custody of children see Chapters 283 and 284, *Acts, 1846–47,* 233.

21. For a discussion of the arguments for equity and for equality see Norma Basch, "Equity vs. Equality: Emerging Concepts of Women's Political Status in the Age of Jackson," *Journal of the Early Republic* 3 (Fall 1983):297–318.

22. Chapter 23, *Acts Passed at a General Assembly of the Commonwealth of Virginia, 1826* (Richmond, 1827), 21.

23. Ibid.

24. Ibid., 22.

25. Ibid.

26. Chapter 71, *Acts of the General Assembly of Virginia, 1840–41* (Richmond, 1841), 78–79. The ambiguous phrase "ecclesiastical law" presumably referred to such cases as those involving consanguinity or coercion of an under-age person. Records of legislative debates and rough bills fail to illuminate the issue any further. The phrase was not used in earlier or later divorce statutes.

27. Chapter 80, *Acts of the General Assembly of Virginia, 1842–43* (Richmond, 1843), 55–56.

28. Chapter 122, *Acts of the General Assembly of Virginia, 1847–48* (Richmond, 1848), 165–67.

29. For a description of views of marriage and family in Virginia see Dickson D. Bruce, Jr., *The Rhetoric of Conservatism: The Virginia Convention of 1829–30 and the Conservative Tradition in the South* (San Marino, Calif., 1982), esp. 141–69; and Jan Lewis, *The Pursuit of Happiness: Family and Values in Jefferson's Virginia* (Cambridge, 1983), 169–208.

30. Chapter 28, *Acts of the General Assembly of Virginia, Passed in 1852–53* (Richmond, 1853), 47–48. Speculation on why legislatures broadened

divorce laws is found in Michael S. Hindus and Lynne E. Withey, "The Law of Husband and Wife in Nineteenth-Century America: Changing Views of Divorce," in D. Kelly Weisberg, ed., *Women and the Law: The Social Historical Perspective*, 2 vols. (Cambridge, Mass., 1982), II, 133–53.

31. Martin J. Schultz, "The Evolution of Marital Disruption in America: A Sociological Examination," Ph.D. dissertation, Southern Illinois University, Carbondale, June 1980, 71–79.

32. Quoted in Censer, " 'Smiling Through Her Tears,' " 27–29.

33. Letters from Charlotte Cullen to Samuel Taylor, "My Dear Sir," Dec. 9 and 10, 1844, in Charlotte Howard Cullen, Letters and Statements concerning her Marital Troubles with John Cullen, M.D., Richmond, Virginia, 1844–46, Brock Collection, Huntington Library, San Marino, Calif.

34. Marylynn Salmon, "Women and Property in South Carolina: The Evidence from Marriage Settlements, 1730 to 1830," *William and Mary Quarterly* 39, 4 (Oct. 1982):655–85; and *Women and the Law of Property in Early America* (Chapel Hill, 1986), 81–119.

35. Marriage Articles of Henry M. Armistead and Mary Robinson, negotiated June 5, 1824, filed Aug. 4, 1824, Clerk's Office, Henrico County, Virginia, Brock Collection, Huntington Library.

36. Marriage Contract between St. Julien de Tournillon and Mary Brown Jones, Parish of East Baton Rouge, July 23, 1832, in Kuntz Collection, Tulane University, New Orleans; Marriage Agreement of Joseph Ozemme Metoyer and Catherine David, Parish of Natchitoches, July 29, 1833; and Marriage Contract between Meric Rachal and Marie Louise Theotisse Metoyer, no parish given, Aug. 27, 1851, both in the Cane River Collection, 1817–71, Historic New Orleans Archives, New Orleans. See also Henry P. Dart, "Marriage Contracts of French Colonial Louisiana," *The Louisiana Historical Quarterly* 17 (April 1934):229–41, for historical background on French marriage contracts.

37. Stanley Clisby Arthur, "Foreword," in Joseph Xavier Delfau de Pontalba; and letter to "Madame," Nov. 12, 1834, from M. Dupoux writing on behalf of Madame de Pontalba, both in the Letters of Baron Joseph X. Pontalba to His Wife, 1796, Special Collections, Tulane University Library, New Orleans.

38. Carl B. Swisher, *The Taney Period, 1836–64*, Vol. V, in *History of the Supreme Court of the United States* (New York, 1974), 756–72.

39. For examples of early legislative divorces see *Acts of the Nineteenth General Assembly of the State of New Jersey*, chap. 102: 958 and chap. 105:959. For dates of prohibition of legislative divorce see Nelson M. Blake, *The Road to Reno: A History of Divorce in the United States* (New York, 1962), 57, 60–61.

40. Early cases are described in Thomas M. Meehan, " 'Not Made Out of Levity': Evolution of Divorce in Early Pennsylvania," *Pennsylvania Magazine of History and Biography* 92 (1968):446–51.

41. *The Statutes at Large of Pennsylvania from 1682–1801* (Harrisburg, 1896–1908), XII, 94; and Meehan, " 'Not Made Out of Levity,' " 455–56.

42. Anonymous, *An Essay on Marriage; or, The Lawfulness of Divorce in Certain Cases Considered* (Philadelphia, 1788), 3, 16, 25.

43. Blake, *Road to Reno*, 57.

44. "An Act for Regulating Marriage and Divorce," *Acts and Resolves of Massachusetts, 1784–1785* (Boston, *ca.* 1892), 564–67.

45. *Acts of the Nineteenth General Assembly of the State of New Jersey* (Trenton, 1795), chap. 114 (Dec. 2, 1794):968–69; and Howard, *Matrimonial Institutions*, II, 11, 13–14.

46. Betty Bandel, "What the Good Laws of Man Hath Put Asunder . . .," *Vermont History* 46 (1978):221–23.

47. Blake, *Road to Reno*, 60–61.

48. *Columbian Register* (New Haven), May 26 and June 9, 1849.

49. *Niles' National Register* 75, 16 (April 18, 1849), 251–52.

50. *The Statutes at Large of Pennsylvania from 1682 to 1801* (Harrisburg, 1896–1908) XII, 96–97; Salmon, *Women and the Law of Property* 66–67, 69–70; and Norma Basch, *In the Eyes of the Law: Women, Marriage and Property in Nineteenth-Century New York* (Ithaca, 1982), 95–99.

51. Alexis de Tocqueville, *Democracy in America* (New York, 1838), 273, 276, 285.

52. *The Statute Laws of the Territory of Iowa, 1838–39* (Dubuque, 1839), 189–91; *Revised Statutes of the Territory of Iowa, 1842–43* (Iowa City, 1843), 169–71; *Acts and Resolutions Passed at the Several Sessions of the Territorial Legislature of Iowa, 1840–46* (Des Moines, 1912), 192; and *Laws of Iowa Passed at the Annual Session of the Legislative Assembly, 1845* (Iowa City, 1846), 659.

53. *Statute Laws of the Territory of Iowa, 1838–39*, 190; and *Revised Statutes of the Territory of Iowa, 1842–43*, 170.

54. Ruth A. Gallaher, *Legal and Political Status of Women in Iowa* (Iowa City, 1918), 69–73.

55. *Laws Passed by the Fourth General Assembly of the State of Illinois, 1824–1825* (Vandalia, 1825), 169.

56. Ibid., 120; and Gallaher, *Legal and Political Status of Women*, 76.

57. Marlin J. Schultz, "The Evolution of Marital Disruption in America," 81–86.

58. For changes in women's lives during the early national era see esp. Ruth H. Bloch, "American Feminine Ideals in Transition: The Rise of the Moral Mother, 1785–1815," *Feminist Studies* 4 (June 1978):101–26; Linda K. Kerber, *Women of the Republic: Intellect and Ideology in Revolutionary America* (Chapel Hill, 1980); Jan Lewis, "The Republican Wife: Virtue and Seduction in the Early Republic," *William and Mary Quarterly* 44 (Oct. 1987):689–721; Karen K. List, "Magazine Portrayals of Women's Role in the New Republic," *Journalism History* 13 (Summer 1986):64–70; Jean V. Matthews, " 'Woman's Place' and the Search for Identity in Ante-Bellum America," *The Canadian Review of American Studies* 10 (Winter 1979):189–304; Mary Beth Norton, *Liberty's Daughters: The Revolutionary Experience of American Women, 1750–1800* (Boston, 1980); and Robert K. Weis, " 'To Please and Instruct the Children,' " *Essex Institute Historical Collections* 123 (April 1987):117–49. For a discussion of single women see Lee Chambers-Schiller, "The Single Woman: Family and Vocation Among Nineteenth-Century Reformers," in Mary Kelley, ed., *Woman's Being, Woman's Place*

(Boston, 1979), 334–50; and *Liberty, A Better Husband: Single Women in America. The Generations of 1780–1840* (New Haven, 1984).

59. *Whisler v. Whisler,* Tipton Circuit Court, Indiana, filed May term, 1845. Court and family records are courtesy of Ellen K. Davison of Kirksville, Mo.

60. Lavina Whisler Estate, Tipton County Court, Indiana, Aug. 14, 1850. Records courtesy of Ellen K. Davison.

61. See for example *Laws of Massachusetts* (Boston, 1821), 508–9.

62. Quoted in Ethel Armes, ed., *Nancy Shippen: Her Journal Book* (New York, 1968), 157.

63. For descriptions of the Kemble/Butler marriage and divorce see J.C. Furnas, *Fanny Kemble: Leading Lady of the Nineteenth-Century Stage* (New York, 1982), 340–50; and Dorothy Marshall, *Fanny Kemble* (New York, 1977), 96–120, 159–222. That Frances Kemble continued to see her children after her return to England is indicated in Fanny Kemble, letter to "My dear Lady Theresa," undated, in Clifton Waller Barrett Collection, University of Virginia Library, Charlottesville.

64. Name of writer illegible, to "My Dear Sister," Dec. 14, 1830, in the Dunlap Family Papers, Tulane University Library Special Collections, New Orleans.

65. Censer, " 'Smiling Through Her Tears,' " 43–45; Lawrence B. Goodheart, Neil Hanks, and Elizabeth Johnson, " 'An Act for the Relief of Females . . .': Divorce and the Changing Legal Status of Women in Tennessee, 1796–1860," Part I, *Tennessee Historical Quarterly* 44 (Fall 1985):334–35; Steven Mintz and Susan Kellogg, *Domestic Revolutions: A Social History of American Family Life* (New York, 1987):60–62; and Michael Grossberg, "Who Gets the Child? Custody, Guardianship, and the Rise of a Judicial Patriarchy in Nineteenth-Century America," *Feminist Studies* 9 (Summer 1983):235–46; and *Governing the Hearth: Law and Family in Nineteenth-Century America* (Chapel Hill, 1985), 238–43.

66. Schultz, "The Evolution of Marital Dissolution in America," 97–98. Tilghman is quoted in Grossberg, *Governing the Hearth,* 239.

67. The process of examining state-by-state divorce provisions and their effect on women has begun. See Norma Basch, "Relief in the Premises: Divorce as a Woman's Remedy in New York and Indiana, 1815–1870," *Law and History Review* (forthcoming, 1991).

68. For New York see Norma Basch, *In the Eyes of the Law: Women, Marriage and Property in Nineteenth-Century New York* (Ithaca, 1982), 94–95; and Blake, *Road to Reno,* 64–79; for Virginia, see Suzanne Lebsock, *The Free Women of Petersburg: Status and Culture in a Southern Town, 1784–1860* (New York, 1984), 34–35, 68–71; and for Louisiana, Dolores Egger Labbe, "Women in Early Nineteenth-Century Louisiana," Ph.D. dissertation, University of Delaware, 66–67.

69. Allen Horstman, *Victorian Divorce* (New York, 1985), 16, 21–22. See also Sybil Wolfram, "Divorce in England, 1700–1857," *Oxford Journal of Legal Studies* 5 (Summer 1985):155–86.

70. Seymour Martin Lipset, ed., *Society in America* by Harriet Martineau (New York, 1962, orig. pub. in 1837), 298–99.

71. Quaker Marriage Certificates, 1776–1808, Brock Collection, Huntington Library, San Marino, Calif.

72. Nancy F. Cott, "Divorce and the Changing Status of Women in Eighteenth-Century Massachusetts," *William and Mary Quarterly* 33 (1976):613; Barbara E. Lacey, "Women in the Era of the American Revolution: The Case of Norwich, Connecticut," *The New England Quarterly* 53 (Dec. 1980):536; and William Nelson, ed., *Documents Relating to the Colonial History of the State of New Jersey* (Paterson, 1900), XXII, cxxiv. In New Jersey, 259 people successfully petitioned for divorce between 1776 and 1844.

73. Benjamin Trumbull, *An Appeal to the Public, Especially the Learned on the Unlawfulness of Divorce* (New Haven, 1788), 36.

74. Zephaniah Swift, *A System of the Laws of the State of Connecticut* (Windham, 1795), I, 192.

75. Meehan, " 'Not Made Out of Levity,' " 163–64.

76. Orville V. Burton, *In My Father's House Are Many Mansions: Family and Community in Edgefield, South Carolina* (Chapel Hill, 1985), 157; Herbert G. Gutman, *The Black Family in Slavery and Freedom, 1750–1925* (New York, 1976), 158; and Deborah Gray White, *Ar'n't I a Woman?* (New York, 1985), 156–57. See also Jo Ann Manfra and Robert R. Dykstra, "Serial Marriage and the Origins of the Black Stepfamily: The Rowanty Evidence," *Journal of American History* 72 (June 1985):18–44.

77. Meehan, " 'Not Made Out of Levity,' " 446; Johnson, *Ante-Bellum North Carolina* 217; Lawrence B. Goodheart, Neil Hanks, and Elizabeth Johnson, " 'An Act for the Relief of Females . . .', Part II, *Tennessee Historical Quarterly* 44 (Winter 1985):402.

78. Susan E. Klepp and Billy G. Smith, "The Records of Gloria Dei Church: Marriages and 'Remarkable Occurrences,' 1794–1806," *Pennsylvania History* 53 (April 1986):125, 132.

79. Salmon, *Women and the Law of Property,* 75–76; Michael S. Hindus and Lynne E. Withey, "The Law of Husband and Wife in Nineteenth-Century America: Changing Views of Divorce," in D. Kelly Weisberg, ed., *Women and the Law: A Social Historical Perspective* (Cambridge, Mass., 1982), 140–41; and quoted in Johnson, *Ante-Bellum North Carolina,* 219–20.

80. *Missouri Intelligencer* (Franklin), June 19, 1824.

81. Lawrence M. Friedman, "Rights of Passage: Divorce Law in Historical Perspective," *Oregon Law Review* 63 (1984):653–54.

82. Abigail Adams, *Letters of Mrs. Adams, Wife of John Adams* (Boston, 1848), 75, 402.

83. *Smith v. Smith,* Feb. 7, 1805, Providence Superior Court Records, Providence, R. I.

84. "On Matrimonial Obedience," *Lady's Magazine* I (July 1792):64–67; and anonymous, "Xantippe," *Godey's Lady's Book* 2 (June 1831):285. Some men also argued that spouses must respect each other. In 1848, a male writer exhorted his readers to view wives as equal and capable beings. See George W. Burnap, *The Sphere and Duties of Woman* (Baltimore, 1848), 193.

85. Grace Greenwood, "Heart Histories," *Godey's Lady's Book* 37 (Au-

gust 1848):72–79. For other writers who supported the idea of woman as a respected companion, see anonymous, "On the Female Character," *Athenaeum*, Series 3, Vol. 5 (Dec. 15, 1830):271–75; anonymous, "Woman," *Dial* 1 (Jan. 1841):362–66; and anonymous, "The True Dignity of the Female Character," *Ladies Repository* 5 (July 1845):194–96.

86. T. S. Arthur, "Ruling a Wife," in *Married Life: Its Shadows and Sunshine* (Philadelphia, 1852), 25–79.

87. Sarah Grimké, *Letters on the Equality of the Sexes* (Boston, 1838), 15, 23, 45, 49; and Margaret Fuller, "The Great Lawsuit," *Dial* 4 (July 1843):28–32. See also Margaret Fuller Ossoli, *Woman in the Nineteenth Century* (New York, 1845).

88. Kerber, *Women of the Republic*, 163. That women also took more control of marital sex during this period is argued in Nancy F. Cott, "Passionlessness: An Interpretation of Victorian Sexual Ideology, 1790–1850," *Signs* 4 (1978):219–36; and Howard Gadlin, "Private Lives and Public Order: A Critical View of the History of Intimate Relations in the U.S.," *Massachusetts Review* 17 (Summer 1976):304–30.

89. *Gentleman's and Lady's Town and Country Magazine*, Sept. 1784, p. 194; and T. S. Arthur, *Sweethearts and Wives; or, Before and After Marriage* (New York, 1843), 53, 163.

90. Robert Dale Owen and Mary Robinson, April 12, 1832; and Henry B. Blackwell and Lucy Stone, 1855, in Aileen S. Kraditor, ed., *Up from the Pedestal: Selected Writings in the History of American Feminism* (Chicago, 1968), 148–50.

91. Ellen K. Rothman, *Hands and Hearts: The History of Courtship in America* (New York, 1984), 24–35; and Karen Lystra, *Searching the Heart: Women, Men, and Romantic Love in Nineteenth-Century America* (New York, 1989), esp. 28–55.

92. Kerber, *Women of the Republic*, 176; Nancy F. Cott, "Eighteenth-Century Family and Social Life Revealed in Massachusetts Divorce Records," *Journal of Southern History* 10 (1976):32; and Johnson, *Ante-Bellum North Carolina*, 202.

93. Phileleutheres, "On the Freedom of Choice, respecting Matrimony," *Boston Magazine* 24 (Oct. 1785):377–79; and anonymous, "The Nuptials," *Godey's Lady's Book* 2 (April 1831):217.

94. T. S. Arthur, *The Stolen Wife: An American Romance* (Philadelphia, 1843), 5–7, 11–14, 30–31; and *Married and Single; or, Marriage and Celibacy Contrasted* (Philadelphia, 1843), 18. See also Herman R. Lantz, Raymond L. Schmitt, and Richard Herman, "The Preindustrial Family in America: A Further Examination of Early Magazines," *American Journal of Sociology* 79 (Nov. 1973):566–88; and Herman R. Lantz, Jane Keyes, and Martin Schultz, "The American Family in the Preindustrial Period: From Base Lines in History to Change," *American Sociological Review* 40 (Feb. 1975):21–36.

95. *The Ladies Magazine*, Sept. 1792, p. 171.

96. Anonymous, "Woman as Wife," *Ladies Repository* 3 (Feb. 1843):53–54. See also anonymous, "Woman's True Glory," *Knickerbocker Magazine* 29 (Jan. 1847):5 and Lydia Sigourney, *Whisper to a Bride* (Hartford, 1850).

97. Lydia Sigourney, *Letters to Mothers* (New York, 1838), viii, 9–10; and Margaret Coxe, *Claims of the Country on American Females* (Columbus, Ohio, 1842), 37. See also anonymous, "Female Influence," *Ladies Repository* 4 (Oct. 1844), 312–13.

98. Nathaniel Hawthorne, *The Scarlet Letter* (Cambridge, Mass., 1883) 311.

99. Trumbull, *Appeal to the Public;* and Timothy Dwight, *Theology: Explained and Defended in a Series of Sermons* (New York, 1828), III, 427.

100. Drake, ed., "A Discourse on Divorce," 435–37.

101. Quoted in Alice S. Rossi, ed., *The Feminist Papers* (New York, 1974), 93. See also Celia Morris Eckhardt, *Fanny Wright: Rebel in America* Cambridge, Mass., 1984), 156, 146.

102. Emma C. Embury, "The Mistaken Choice, or, Three Years of Married Life," *Grahm's Magazine* 19 (July 1841):13–17. Other ways to deal with unhappy marriage was to avoid it altogether—see anonymous, "Matrimony," *Knickerbocker Magazine* 16 (Nov. 1840):390–93; and Enna Duval, "Old Maids," *Grahm's Magazine* 30 (March 1847):193–99—or to escape it by committing suicide—see Robert Morris, "The Neglected Wife," *Grahm's Magazine* 19 (Aug. 1841):58–60.

103. E.D.E.N. Southworth, *The Deserted Wife* (New York, 1850), 5, 7.

104. Ibid., 117–23, 153, 176.

105. T. S. Arthur, *The Divorced Wife* (Philadelphia, 1850), 7–24, 53, 80–96.

106. That divorce was in place before urbanization and industrialization occurred and that these two factors exercised less impact on divorce than is usually thought, see Alex Inkeles, "The Responsiveness of Family Patterns to Economic Change in the United States," *The Tocqueville Review* 6 (Spring–Summer 1984):5–50; Herman Lantz, Martin Schultz, Mary O'Hara, "The Changing American Family from the Preindustrial to the Industrial Period: A Final Report," *American Sociological Review* 42 (June 1977):406–21; and Schultz, "The Evolution of Marital Disruption in America," 78–85, 89, 92, 105–6. For a description of the Jackson marriage see Harriet Chappell Owsley, "The Marriages of Rachel Donelson," *Tennessee Historical Quarterly* 36 (Winter 1977):479–92.

**Chapter 3**

1. *New York Tribune*, Dec. 18, 1852.

2. Ibid., March 1, 1860.

3. Nelson Manfred Blake, *The Road to Reno: A History of Divorce in the United States* (New York, 1962), 116–29; and Mary Somerville Jones, "An Historical Geography of Changing Divorce Laws in the United States," Ph.D. dissertation, University of North Carolina, Chapel Hill, 1978, pp. 23–34.

4. *Report of the Debates and Proceedings of the Convention for the Revision of the Constitution of the State of Indiana, 1850* (Indianapolis, 1850), I, 40,

58; II, 1274, 1279; and *Journal of the Convention of the People of the State of Indiana, to Amend the Constitution, 1850* (Indianapolis, 1851), 26, 169, 506, 514, 967. For a discussion of legislative divorce in the Midwest see George E. Howard, *A History of Matrimonial Institutions* 3 vols. (Chicago, 1904), III, 96–98.

5. *New York Tribune,* March 5, 1860. For discussion of the full debate see Richard Wires, "The Greeley-Owen Divorce Debate of 1860," *Forum* III (Spring 1962):49–60.

6. For 1852 divorce statutes see *The Revised Statutes of the State of Indiana, 1852* (Indianapolis, 1852), II, 233–38. See also Richard Wires, *The Divorce Issue and Reform in Nineteenth-Century Indiana* (Muncie, 1967), 5–6, 14; and Val Nolan, Jr., "Indiana: Birthplace of Migratory Divorce," *Indiana Law Journal* 26 (Summer 1951):517–18.

7. For example, according to one source, New Jersey adopted the "notification by publication" provision as early as March 4, 1795. William Nelson, ed., *Documents Relating to the Colonial History of the State of New Jersey* (Paterson, 1900), XXII, cxxv. For an example of such a notice from a western state besides Indiana see "Notice by attorney of Amelia P. Campbell to her husband David H. Campbell who cannot be found in county or state," *Missouri [Franklin] Intelligencer,* May 12, 1826.

8. Wires, *The Divorce Issue and Reform,* 16–17.

9. Wires, *The Divorce Issue,* 20–32; and *Indianapolis Daily Journal,* Dec. 3, 1858.

10. *Laws of the State of Indiana, 1859* (Indianapolis, 1859), 108–10.

11. Auguste Carlier, *Marriage in the United States* (New York, 1867), 111–13.

12. Eli Lilly, editor, *Schliemann in Indianapolis* (Indianapolis, 1961), 8, 12–16, 21–22, 49–50; and David A. Traill, "Schliemann's American Citizenship and Divorce," *The Classical Journal* 77 (April/May 1982):336–40.

13. Lilly, *Schliemann in Indianapolis,* 61–63, 68.

14. Baker is quoted in the *Journal of the Senate, 47th Session* (Indianapolis, 1871), 59–62. See also Wires, *The Divorce Issue,* 25–32; and *Laws of the State of Indiana, Passed at the Forty-Eighth Regular Session of the General Assembly, 1873* (Indianapolis, 1873), 107–12.

15. Quoted in James Harwood Barnett, *Divorce and the American Divorce Novel, 1858–1937* (New York, 1968 ed.), 23; and *Indianapolis Daily Journal,* Dec. 3, 11, 1858.

16. William Dean Howells, *A Modern Instance* (Bloomington, 1977 ed.), xv, 398, 404–10, 418, 438. See also George R. Uba, "Status and Contract: The Divorce Dispute of the 'Eighties and Howells's *A Modern Instance,*" *Colby Library Quarterly* 19 (June 1983):78–89.

17. Nolan, "Indiana," 525–26. For other interpretations see Alexander A. Plateris, *100 Years of Marriage and Divorce Statistics, United States, 1867–1967* (Rockville, Md., 1973), 34–35, 59; and Wires, *The Divorce Issue,* 8.

18. Quoted in Raymond Lee Muncy, *Sex and Marriage in Utopian Communities: 19th Century America* (Bloomington, 1973), 203. After Wright's utopian experiment ended in 1828, she abandoned her free love princi-

ples and married. In 1850, however, she divorced her husband in Raleigh, Tennessee.

19. See Muncy, *Sex and Marriage in Utopian Communities,* for a discussion of the views of these and other utopian communities of the nineteenth century.

20. Carlier, *Marriage in the United States,* 98–99.

21. Quoted in Louis J. Kern, *An Ordered Love: Sex Roles and Sexuality in Victorian Utopias* (Chapel Hill, 1981), 121–22.

22. Ibid., 121.

23. Louis J. Kern, "Ideology and Reality: Sexuality and Women's Status in the Oneida Community," *Radical History Review* 20 (Spring/Summer 1979):180–204; and Lawrence Foster, "Free Love and Feminism: John Humphrey Noyes and the Oneida Community," *Journal of the Early Republic* I (Summer 1981):165–83.

24. Oneida Community, *Slavery and Marriage: A Dialogue* (Oneida, 1850), 7–13.

25. Kern, *An Ordered Love,* 273; and Muncy, *Sex and Marriage in Utopian Communities,* 194–95.

26. Anonymous, "Marriage and Divorce," *Southern Quarterly Review* 10 (1854):332–55.

27. The Winget case is discussed in Susan Freda Hult, "Divorce and the Decline of Patriarchy in South Carolina," M.A. thesis, 1985, Clemson University, Clemson, S. C., 12–13; and the other cases in Lawrence T. McDonnell, "Desertion, Divorce and Class Struggle: Contradictions of Patriarchy in Antebellum South Carolina," unpublished paper read at the meetings of the Southern History Association in Houston, Texas, in November 1985, pp. 3–4.

28. Edwin J. Scott, *Random Recollections of a Long Life* (Columbia, S.C., 1884), 35–36, 83–84, 133–35.

29. J. Nelson Frierson, "Divorce in South Carolina," *North Carolina Law Review* 9 (April 1931):271–72.

30. Carlier, *Divorce in the United States* 109–10; and *Code of Laws of South Carolina* (Rochester, 1949), 62.

31. Frierson, "Divorce in South Carolina," 265–66.

32. Orville V. Burton, *In My Father's House Are Many Mansions: Family and Community in Edgefield, South Carolina* (Chapel Hill, 1985), 157, 160, 166–67, 293–95, 316. For more about Hammond see Carol Bleser, ed., *The Hammonds of Redcliffe* (New York, 1981).

33. Joel Prentiss Bishop, *Commentaries on the Law of Marriage and Divorce* (Boston, 1881 ed.), I, 22–26, 30.

34. Barnett, *Divorce and the American Divorce Novel,* 34–35, 37, 45.

35. *New York Tribune,* Dec. 1, 18, 24, 1852, and Jan. 28, 1853. See also anonymous, *Love, Marriage, and Divorce and the Sovereignty of the Individual: A Discussion Between Henry James, Horace Greeley, and Stephen Pearl Andrews* (Boston, 1889).

36. *New York Tribune,* March 1, 5, 12, 1860. The entire text of the debates is found in Horace Greeley, *Recollections of a Busy Life* (New York, 1869), 570–618.

37. Elizabeth Cady Stanton, Susan B. Anthony, and Mathilda J. Gage, *History of Woman Suffrage* (New York, 1881), I, 717, 720.

38. Ibid., 722. For a discussion of Stanton's and other women's rights leaders' views of love and marriage, see William Leach, *True Love and Perfect Union: The Feminist Reform of Sex and Society* (New York, 1980).

39. *The Revolution,* Dec. 23, 1869, p. 385. See also Elisabeth Griffith, "Elizabeth Cady Stanton on Marriage and Divorce: Feminist Theory and Domestic Experience," 233–51, in Mary Kelley, ed., *Woman's Being, Woman's Place* (Boston, 1979); and Elizabeth Clark, "Matrimonial Bonds: Slavery, Contract and the Law of Divorce in Nineteenth-Century America," Legal History Program Working Papers, Series 1, no. 10, June 1987, pp. 1–37, Institute for Legal Studies, University of Wisconsin-Madison Law School. For feminist ideas concerning equitable marriage see Blanche Glassman Hersh, " 'A Partnership of Equals': Feminist Marriages in 19th-Century America," in Elizabeth H. Pleck and Joseph Pleck, *The American Man* (Englewood Cliffs, N.J., 1980).

40. Françoise Basch, "Women's Rights and the Wrongs of Marriage in Mid-Nineteenth-Century America," *History Workshop Journal* 22 (Autumn 1986):28–29

41. Ibid.

42. Doggerel quoted in Jerome Nadelhaft, "Wife Torture: A Known Phenomenon in Nineteenth-Century America," *Journal of American Culture* 10 (Fall 1987):42. For Bloomer and Swisshelm, see Elizabeth H. Pleck, *Domestic Tyranny: The Making of Social Policy Against Family Violence* (New York, 1987), 55–56.

43. *New York Tribune,* Nov. 26, Dec. 1, 3, 1869, and May 11, 1870.

44. *New York Times,* Dec. 4, 1869; and *The Revolution,* Dec. 23, 1869, p. 385.

45. Anonymous, *The Richardson-McFarland Tragedy* (Philadelphia, 1870), 89–90.

46. *The Revolution* V (May 19, 1870):329–30; and Ellen DuBois, "On Labor and Free Love: Two Unpublished Speeches of Elizabeth Cady Stanton," *Signs* I (Autumn 1975):265–68.

47. *The Independent* XXII (May 12, 1870):4 and (May 19, 1870):4.

48. Robert Shapen, *Free Love and Heavenly Sinners: The Story of the Great Henry Ward Beecher Scandal* (New York, 1954), 123–64; Altina L. Waller, *Reverend Beecher and Mrs. Tilton* (Amherst, Mass., 1982), 1–17; and *Blood v. Blood,* Sept. 18, 1876, in Woodhull-Martin Collection, General Correspondence, 1876–80, Morris Library, University of Southern Illinois, Carbondale.

49. Stanton is quoted in Alma Lutz, *Created Equal: A Biography of Elizabeth Cady Stanton* (New York, 1940), 226. For a more detailed account of the mid-century divorce debate see Sidney Ditzion, *Marriage, Morals, and Sex in America: A History of Ideas* (New York, 1969), 125–57.

50. Quoted in Louise R. Noun, *Strong-Minded Women: The Emergence of the Woman-Suffrage Movement in Iowa* (Ames, 1969), 186.

51. Ibid., 188.

52. Ibid., 205.

53. George F. Parker, *Iowa Pioneer Foundations* (Iowa City, 1904), 459–61.

54. Figures cited in Lynne Carol Halem, *Divorce Reform: Changing Legal and Social Perspectives* (New York, 1980), 28.

55. Carlier, *Marriage in the United States*, 114–15.

56. U.S. Dept. of Commerce and Labor, *Marriage and Divorce*, I, 11.

57. Ibid., 24–25.

58. Ibid., 20–21.

59. Ibid., 16–18, 44–45.

60. Ibid., 19; and Sybil Wolfram, "Divorce in England, 1700–1857," *Oxford Journal of Legal Studies* 5 (Summer 1985):182.

61. R.A. Pierson, letter to "Dear Brother" (James F. Pierson), Jan. 14, 1862; and letter to "Dear Sister" (Mary C. Pierson), Jan. 18, 1862, Kuntz Collection, Tulane University, New Orleans; and Dorothea Rhodes Lummis Moore, letter to "Dearie boy" (Charles Fletcher Lummis), April 26 or 27, 1884, in Letters, 1883–84, Huntington Library, San Marino.

62. Karen Lystra, *Searching the Heart: Women, Men, and Romantic Love in Nineteenth-Century America* (New York, 1989), esp. 157–226; Steven Mintz, *A Prison of Expectations: The Family in Victorian Culture* (New York, 1983), 103–46; and Ellen E. Rothman, *Hand and Hearts: The History of Courtship in America* (New York, 1984), 90–114.

63. One example of the call for interdependence and mutual respect argument is Gail Hamilton, pseudonym used by Mary Dodge, *Woman's Wrongs* (Boston, 1869), 133, 183, 199, 203. Beecher's position is found in Catharine Beecher, *The American Woman's Home; or, Principles of Domestic Science* (New York, 1870), 13, 23, 265, 466.

64. T. S. Arthur, *Out in the World* (New York, 1864), 8, 12; and "Hints to Husbands" and "A True Wife," in *Our Homes; Their Cares and Duties, Joys and Sorrows* (Philadelphia, 1858), 264–66. 283.

65. Robert L. Griswold, "Adultery and Divorce in Victorian America," Legal History Program Working Papers, Series 1, No. 6, March 1986, Institute for Legal Studies, University of Wisconsin-Madison Law School, 10. That adultery decreased in relative importance as a ground for divorce and cruelty increased is supported by U.S. Dept. of Commerce and Labor, *Marriage and Divorce, 1867–1906* (Westport, Conn., 1978 reprint ed.), I, 26.

66. For descriptions of wife abuse see Myra C. Glenn, "Wife-Beating: The Darker Side of Victorian Domesticity," *Canadian Review of American Studies* 15 (Spring 1984):17–33; and Nadelhaft, "Wife Torture," 39–59. Quotes are found in U.S. Commissioner of Agriculture, *Annual Report* (Washington, 1862), 462–70; and *Laramie [Wyoming] Sentinel,* Oct. 10, 1885.

67. Robert L. Griswold, "Sexual Cruelty and the Case for Divorce in Victorian America," *Signs* 2 (Spring 1986):529–41. For discussions of sexual beliefs during these years see Carl N. Degler, "What Ought To Be and What Was: Women's Sexuality in the Nineteenth Century," *American Historical Review* 79 (Dec. 1974):1467–90; G.J. Barker-Benfield, "The Spermatic Economy: A Nineteenth-Century View of Sexuality," 370–401, in Michael Gordon, ed., *The American Family in Social and Historical Perspective* (New York, 1978); and Mabel Collins Donnelly, *The American Victorian Woman: The Myth and the Reality* (New York, 1986).

68. Quoted in Robert L. Griswold, "Law, Sex, Cruelty, and Divorce in Victorian America, 1840–1900," *American Quarterly* 38 (Winter 1986):721, 726.

69. E.D.E.N. Southworth, *Ishmael; or, In the Depths* (Philadelphia, 1876), 621; and Mary Grace Halpine, *The Divorced Wife* (1885), 8–9, 68, 86, 149.

70. Michael Grossberg, *Governing the Hearth: Law and the Family in Nineteenth-Century America* (Chapel Hill, 1985), 248.

71. Dorothea Rhodes Lummis Moore, letter to "Dearie boy."

72. Michael Grossberg, "Who Gets the Child? Custody, Guardianship, and the Rise of a Judicial Patriarchy in Nineteenth-Century America,'' *Feminist Studies* 9 (Summer 1983):246–48; and *Governing the Hearth*, 244–47, 72.

## Chapter 4

1. *Missouri [Franklin] Intelligencer*, June 19, 1824.

2. Albert D. Richardson, *Beyond the Mississippi* (Hartford, 1867), 148. That such factors as age and occupation led to higher divorce rates in the West is argued in Henry Pang, "Highest Divorce Rates in Western United States," *Sociology and Social Research* 52 (Jan. 1968):228–36. For the effect of high mobility on western divorce rates see Bill Fenelon, "State Variations in United States Divorce Rates," *Journal of Marriage and the Family* 33 (May 1971):321–27. For the argument that the western environment had a similar liberalizing effect on women's rights see Mari J. Matsuda, "The West and the Legal Status of Women: Explanations of Frontier Feminism," *Journal of the West* 24 (Jan. 1985):47–56.

3. U.S. Dept. of Commerce and Labor, *Marriage and Divorce, 1867–1906*, 2 vols. (Westport, Conn., 1978, reprint ed.), I, 14–15, 70–71. See also Mary Somerville Jones, "An Historical Geography of Changing Divorce Law in the United States," Ph.D. dissertation, University of North Carolina, Chapel Hill, 1978, pp. 59–60.

4. U.S. Dept. of Commerce and Labor, *Marriage and Divorce, 1867–1906*, I, 14–15.

5. Ibid., I, 25, 86, 88, 90, 92.

6. Miriam M. Worthington, ed., "Diary of Anna R. Morrison, Wife of Isaac L. Morrison," *Illinois State Historical Society Journal* 7 (April/Jan., 1914–15):34–50; and Case #1966, *Hansen v. Hansen*, filed July 25, 1895, Territorial Records, Logan County, Guthrie, Okla. A description of desertion cases in San Diego between 1850 and 1880 are found in Susan Gonda, "San Diego Women and Frontier Divorce," undated, San Diego Historical Society.

7. Case #1945, *Ball v. Ball*, filed June 21, 1895, Territorial Records, Logan County, Guthrie.

8. Caroline Phelps, Diary, 1830–1860, Iowa State Historical Society, Iowa City.

9. Robert Glass Cleland, ed., *Apron Full of Gold: The Letters of Mary Jane Megquier* (San Marino, 1949), 87.

10. Walter L. Davis Poster, 1905, Police Record, Criminals Wanted, 1906, Oklahoma Territorial Museum, Guthrie.

11. George H. Blowers Poster, 1905, Criminals Wanted, 1906, Oklahoma Territorial Museum.

12. Robert Archibald Poster, Peter J. Hempler Poster, "Disappeared— Frank Limbrack" Poster, and "A Liberal Reward—George W. Roach" Poster, all 1906, Criminals Wanted, 1906, Oklahoma Territorial Museum.

13. Mabel Wade Poster, 1905, and "Fifty Dollars Reward" Poster by S. D. Gilbert, 1907, Criminals Wanted, 1906, Oklahoma Territorial Museum.

14. U.S. Dept. of Commerce and Labor, *Marriage and Divorce*, I, 25–26, 86, 88, 90, 92.

15. For a discussion of the cruelty phenomenon in Montana, see Paul Petrik, "If She Be Content: The Development of Montana Divorce Law, 1865–1907," *Western Historical Quarterly* 23 (July 1987):261–92; and *No Step Backward: Women and Family on the Rocky Mountain Mining Frontier, Helena, Montana 1865–1900* (Helena, 1987), 97–114.

16. Cases #10, *Ames v. Ames,* filed July 20, 1890; #725; *Wilder v. Wilder,* May 10, 1892; #1958; *Winston v. Winston,* August [no day given] 1895; #2290, *Brown v. Brown,* Jan. 19, 1897, all in Territorial Records, Logan County, Guthrie. In the Ames and Brown cases, the paperwork ended unaccountably. Wilder and Winston, however, received divorce decrees. A description of cruelty cases in San Diego between 1850 and 1880 is found in Gonda, "San Diego Women and Frontier Divorce."

17. Case #1978, *Thorpe v. Thorpe,* filed July 25, 1895, Territorial Records, Logan County. Similar findings in California cases between 1850 and 1890 are reported in Robert L. Griswold, *Family and Divorce in California: Victorian Illusions and Everyday Realities, 1850–1890* (Albany, N. Y., 1982), esp. 1–17, 120–40, 170–79.

18. U.S. Dept. of Commerce and Labor, *Marriage and Divorce*, I, 24.

19. Ibid., II, 79–164. The northeastern states came remarkably close to the national profile except in Maine, Massachusetts, and Rhode Island, where women got slightly more than two-thirds of the divorces. In southern states, men came much closer to women, or exceeded them, in numbers of divorces they obtained. Between 1867 and 1906, men got more divorces than women in the states of Alabama, Florida, Georgia, Mississippi, North Carolina, and Virginia.

20. U.S. Dept. of Commerce and Labor, *Marriage and Divorce*, I, 33, 99.

21. Case #1406, *Lyman v. Lyman,* filed Aug. 30, 1891; and Case #2387, *Hughes v. Hughes,* Aug. 25, 1897.

22. Cases #714, *Gleason v. Gleason,* filed July 2, 1891; 736, *Weyach v. Weyach,* July 29, 1891; 798, *Zeller v. Zeller,* Oct. 18, 1891; 1399, *Quick v. Quick,* Aug. 23, 1893; 1410, *Wise v. Wise,* Aug. 26, 1893; 1944, *Rowland v. Rowland,* May 28, 1895; 1984, *Schwart v. Schwart,* Aug. 3, 1895; 1994, *Beck v. Beck,* Aug. 12, 1895; 2681, *Bowers v. Bowers,* Jan. 17, 1899; and 2702, *Condron v. Condron,* Feb. 15, 1889, Territorial Records, Logan County. Similar findings are described in Robert L. Griswold, "Apart But

Not Adrift: Wives, Divorce, and Independence in California, 1850–1890,"
*Pacific Historical Review* 49 (May 1980):165–87.

23. Paul Frison, The Life of Martha Waln, Pioneer of Tensleep, Wyoming, 1939, Wyoming State Archives and Museum, Cheyenne. The Hastings woman is described in Luna Kellie, Memoirs, undated, Nebraska State Historical Society, Lincoln; and the woman homesteader in Paula M. Bauman, "Single Women Homesteaders in Wyoming, 1880–1930," *Annals of Wyoming* 58 (Spring 1986):42. Wives who escaped marriage by working as domestics and seamstresses are found in Police Record, Criminals Wanted, 1906, Guthrie. The piano teacher is described in Brett Harvey Vuolo, "Pioneer Diaries: The Untold Story of the West," *Ms. Magazine* 3 (May 1975):32–34. The woman who took over the farm is Maranda J. Cline, Diary, 1891–1907, Iowa State Historical Society, Iowa City. The woman with eight sons is found in James Elder Armstrong, *Life of a Woman Pioneer* (Chicago, 1931).

24. U.S. Dept. of Commerce and Labor, *Marriage and Divorce*, II, 608–57. In northeastern states, men got custody of their children approximately one-third of the time. In southern states, where men got more of the divorces, they exceeded women in custody awards in the states of Alabama, Mississippi, North Carolina, and Virginia.

25. *Shawnee [Okla.] News*, Aug. 2, 3, 4, Sept. 21, Oct. 31, Nov. 1, 1906. For a sketch of Benjamin Blakeney see Luther B. Hill, *A History of the State of Oklahoma* (Chicago, 1909), 16–17.

26. Dorothy Florence Towe (Lena's husband changed Towe to Tow) Tester, "The Circle Complete," in Montana American Mothers Bicentennial Project, 1975–76, Montana Historical Society, Helena; and Martha Farnsworth, Diaries, 1890–93, Kansas State Historical Society, Topeka. Fiction also revealed that women stuck to husbands in silent submission. See Robert H. Solomon, "The Prairie Mermaid: Love-Tests of Pioneer Women," *Great Plains Quarterly* 4 (Summer 1984):143–51.

27. *The Daily [Guthrie] Oklahoma State Capital*, July 6, 1893; and *The Vinita [Okla.] Chieftain*, Sept. 1, 1904.

28. "White Woman Asks Divorce from a Negro," *Oklahoma City Times-Journal*, July 6, 1908.

29. For an account of early divorce practices among the Cherokee see John Phillip Reid, *A Law of Blood: The Primitive Law of the Cherokee Nation* (New York, 1970), 117–18; and Renard Strickland, *Fire and the Spirits: Cherokee Law from Clan to Court* (Norman, 1975), 97–102. Examples of divorce decrees are found in Cherokee National Council, Vol. 270, Nov. 13, 1880, pp. 232–33; and Vol. 279, Nov. 22, 1882, p. 14, Archives and Manuscript Division, Oklahoma Museum and Historical Society, Oklahoma City; and *Laws and Joint Resolutions of the Cherokee Nation* (Tahlequah, Cherokee Nation, 1887), 81; and are reported in *The Cherokee Advocate* (Tahlequah, Indian Territory), Nov. 16, 1880, Sept. 29, Dec. 1, 8, 15, 1882, and Feb. 16, 1883. Statutes are found in *Compiled Laws of the Cherokee Nation* (Tahlequah, Indian Territory, 1881), 287. Other tribes practiced divorce as well. See for example *Constitution, Laws, and Treaties of the Chickasaws* (Tishomingo City, Okla., 1860), 102–4; *Constitution, Laws and Treaties of*

the *Chickasaws* (Sedalia, Mo., 1878), 67–69; and *General and Special Laws of the Chickasaw Nation* (Muskogee, Okla., 1884), 19. For divorce statutes in the Indian Territory see *Annotated Statutes of the Indian Territory* (St. Paul, Minn., 1899), 324–27.

30. Luke Bearshield's divorce was reported in *The [Guthrie] Daily Leader,* July 14, 1893, and *The [Oklahoma City] Oklahoma Times Journal,* July 16, 1893. Cinda Richard's divorce petition was reported in the *Daily Oklahoman [Oklahoma City],* Sept. 28, 1913. Reports of other Indian divorces are found in *The [Guthrie] Daily Leader,* April 25, 1894; *The [Vinita] Indian Chieftain,* Nov. 26, 1896; and *The Vinita [Oklahoma] Weekly Chieftain,* Oct. 20, 1904.

31. U. S. Dept. of Commerce and Labor, *Marriage and Divorce,* I, 45–46. Newspaper reports of divorces are found in *The [Oklahoma City] Oklahoma Times Journal,* Sept. 22, 1893; *The [Guthrie] Daily Leader,* Nov. 24, 1893; *The [Oklahoma City] Times Journal,* April 4, 1894; *The Daily [Guthrie] Oklahoma State Capital,* Dec. 19, 1894; *The Daily Times Journal [Oklahoma City],* April 15, 1895; and *The Weekly Chieftain [Vinita, Okla.],* Nov. 19, 1909. Divorce even touched western governors. See Constance Wynn Altshuler, "The Scandalous Divorce: Governor Safford Severs the Tie That Binds," *The Journal of Arizona History* 30 (Summer 1989):181–92. That divorce affected all classes of Californians is demonstrated in Griswold, *Family and Divorce in California,* 25–26, 180–81.

32. For an example of a sensationalized divorce trial see Donald W. Hamblin, "The Sharon Cases: A Legal Melodrama of the Eighties," *Los Angeles Bar Bulletin* 25 (Dec. 1949):101–3, 122–23, 125–28. Another description of the Sharon trial is found in Edna Byran Buckbee, "The Story of Sarah Althea Hill," *The Pony Express* 20 (March 1954):4, 15. Another controversial California divorce case is described in William M. Kramer and Norton B. Stern, "An Issue of Jewish Marriage and Divorce in San Francisco," *Western States Jewish History* 21 (Oct. 1988):46–57.

33. Samuel W. Dike, *Important Features of the Divorce Question* (Royalton, Vt., 1885).

34. U.S. Dept. of Commerce and Labor, *Marriage and Divorce,* I, 22; and Hugo Hirsch, *Hirsch's Tabulated Digest of the Divorce Laws of the United States* (Brooklyn, 1888), unpaged chart.

35. Margaret Lee, *Divorce; or, Faithful and Unfaithful* (New York, 1889), 5–8.

36. For a detailed history of the Latter-day Saints see James B. Allen and Glen M. Leonard, *The Story of the Latter-day Saints* (Salt Lake City, 1976). Women's accounts are found in Kenneth W. Godfrey, Audrey M. Godfrey, and Jill Mulvay Derr, *Women's Voices: An Untold History of the Latter-day Saints, 1830–1900* (Salt Lake City, 1982).

37. John Gunnison, letter to "My Dear Martha," March 1, 1850; and Philip Taylor, letter to "My Dear Sir," Jan. 22, 1879, both at the Huntington Library, San Marino. For a fuller description of the reaction to Mormon polygamy, see Carol Weisbrod and Pamela Sheingorn, "*Reynolds v. United States:* Nineteenth-Century Forms of Marriage and the Status of Women," *Connecticut Law Review* 10 (Summer 1978):828–58; Charles A.

Cannon, "The Awesome Power of Sex: The Polemical Campaign Against Mormon Polygamy," in Thomas L. Altherr, ed., *Procreation or Pleasure?: Sexual Attitudes in American History* (Malabar, Fla., 1983), 99–113. An attack on polygamy's effect on women is Jennie Anderson Froiseth, ed., *The Women of Mormonism; or, The Story of Polygamy as Told by the Victims Themselves* (Detroit, 1882). More recent scholarly accounts of Mormon women's views of polygamy are Julie Dunfey, " 'Living the Principle' of Plural Marriage: Mormon Women, Utopia, and Female Sexuality in the Nineteenth Century," *Feminist Studies* 10 (Fall 1984):523–36; Joan Iversen, "Feminist Implications of Mormon Polygyny," ibid., 505–21; Kahlile Mehr, "Women's Response to Plural Marriage," *Dialogue: A Journal of Mormon Thought* 18 (Fall 1985):84–97; and Jessie L. Embry and Martha S. Bradley, "Mothers and Daughters in Polygamy," ibid., 99–107.

38. Eugene E. Campbell and Bruce L. Campbell, "Divorce among Mormon Polygamists: Extent and Explanations," *Utah Historical Quarterly* 46 (1978):4–23. See also Kimball Young, *Isn't One Wife Enough?* (New York, 1954).

39. Quoted in Lawrence Foster, "Polygamy and the Frontier: Mormon Women in Early Utah," *Utah Historical Quarterly* 50 (Summer 1982):285.

40. *Utah Territorial Laws* (Salt Lake City, 1852), 82–84.

41. Probate Court Records Book, Feb. 12, 1856, Washington County, Utah, Huntington Library, San Marino.

42. Probate Court Records Book, Nov. 24, 1857, Washington County, Utah.

43. Jacob Smith Boreman, Curiosities of Early Utah Legislation, 1905, pp. 40–42, Huntington Library, San Marino. Boreman's notations omit any speculation that loose divorce laws may have been put in place by Mormons to accommodate new converts who wished to dispose of dissenting spouses. If Boreman could have foretold the future, he would have seen that the willingness of Utah courts to avoid adversarial suits anticipated no-fault divorce by about one hundred years. For a fuller discussion of Utah territorial divorce law, see Richard I. Aaron, "Mormon Divorce and the Statute of 1852: Questions for Divorce in the 1980s," *Journal of Contemporary Law* 8 (1982):5–45.

44. Jacob Smith Boreman, Reminiscences of My Life in Utah, On and Off the Bench, 1872–1877, Huntington Library, San Marino. For a fuller discussion of the Young divorce case, see Kern, *An Ordered Love,* 198–200.

45. Aaron, "Mormon Divorce," 22.

46. Doane Robinson, "Divorce in Dakota," *South Dakota Historical Collection* (Pierre, 1924), XII, 268–72; Howard R. Lamar, *Dakota Territory, 1861–1889: A Study of Frontier Politics* (New Haven, 1956), 93; "Sioux Falls: The Origins of an Early Divorce Capital," *Sioux Falls Tribune,* Aug. 17, 1893. For divorce laws, see *Laws, Memorials and Resolutions of the Territory of Dakota, 1865–66* (Yankton, 1865–66), 11–14; *General Laws, Memorials and Resolutions of the Territory of Dakota, 1867* (Yankton, 1867), 45–50; George H. Hand, ed., *The Revised Codes of the Territory of Dakota, 1877* (Yankton, 1880), 215–19; A.B. Levisee and L. Levisee, eds., *The Annotated Revised*

*Codes of the Territory of Dakota, 1883* (St. Paul, 1885), 747–51; and *The Compiled Laws of the Territory of Dakota, 1887* (Bismarck, 1887), 545–51.

47. "Enabling Act," *Laws Passed at the First Session of the Legislative Assembly of the State of North Dakota* (Bismarck, 1890), 3–13.

48. *Evening Argus Leader,* Jan. 10, 1893. See also Myra A. Strasser, "Social and Cultural Development in 19th Century Sioux Falls, South Dakota," M.A. thesis, University of South Dakota, Vermillion, 1969.

49. Charles A. Smith, *History of Minnehaha County, South Dakota* (Mitchell, S.D., 1949), 373; Harry Hazel and S. L. Lewis, *The Divorce Mill: Realistic Sketches of the South Dakota Divorce Colony* (New York, 1895), 5; and George Fitch, "Shuffling Families in Sioux Falls: How a Little Town Has Become a Big City through Its Divorce Industry," *American Magazine* 66 (Sept. 1908):443–45, 448.

50. Will Lillibridge, *The Dissolving Circle* (New York, 1908), 6; and Jane Burr, *Letters of a Dakota Divorce* (Boston, 1909).

51. Bishop William Hobart Hare, "Notes on Women," undated, Papers of Bishop William Hobart Hare, 1864–1909, Center for Western Studies, Augustana College, Sioux Falls, S. D.; and *Missionary District of South Dakota: Journal of the Convocations with the Annual Address of the Bishop* (1890), xi.

52. Bishop William Hobart Hare, *Marriage and Divorce* (Sioux Falls, 1893), 2–3. For a discussion of the pastoral and Hare's position, see M.A. DeWolfe Howe, *The Life and Labors of Bishop Hare: Apostle to the Sioux* (New York, 1911), 354–77. For his fears concerning the school, see Virginia Driving Hawk Sneve, *That They May Have Life: The Episcopal Church in South Dakota, 1859–1976* (New York, 1977), 127.

53. *Sioux Falls Argus-Leader,* Feb. 25, 1895; Sneve, *That They May Have Life,* 128; and "Divorce in South Dakota," *The Nation* 56 (Jan. 26, 1893):60–61.

54. Lonnie R. Haugland, "South Dakota Divorce Legislation and Reform: 1862–1908," *The Region Today: A Quarterly Journal of the Social Science Research Associates* 2 (May 1975):54–55.

55. Mary E. Peabody, *Zitana Duzahon: Swift Bird, The Indians' Bishop* (Hartford, Conn., 1915), 61–63; Haugland, "South Dakota Divorce Legislation," 55–56; and Herbert S. Schell, *History of South Dakota* (Lincoln, 1968), 261–62.

56. The two analysts are Arthur G. Horton, *An Economic and Social Survey of Sioux Falls, South Dakota, 1938–39* (1939), 81; and Robinson, "Divorce in Dakota," 275–76. Department of Labor statistics are found in Alexander A. Plateris, *100 Years of Marriage and Divorce Statistics, United States, 1867–1967* (Rockville, Md., 1973), 34–35. For other statistical interpretations see Nelson M. Blake, *The Road to Reno: A History of Divorce in the United States* (New York, 1962), 123.

57. Lois Abby Lane, "The Divorcees," undated, Old Courthouse Museum, Sioux Falls, S. D.; Smith, *History of Minnehaha County,* 372–74; and *Sioux Falls Tribune* (S. D.), Aug. 24 and 28, 1983.

58. Charles A. Pollock, "The Divorce Law," undated, Pollock Collec-

tion, 1890, 1923–70, North Dakota Institute for Regional Studies, North Dakota State University, Fargo.

59. *Laws Passed at the First Session of the Legislative Assembly of the State of North Dakota* (Bismarck, 1890), 22, 54; and *The Revised Laws of the State of North Dakota, 1899* (Bismarck, 1899), 695.

60. William H. White, "Early History and Settlements of Cass County," undated, North Dakota Institute for Regional Studies, North Dakota State University, Fargo; and Cass County Historical Society, *Rural Cass County: The Land and Its People* (Dallas, 1976), 8.

61. Elizabeth Preston Anderson, "Under the Prairie Winds," undated, in Preston Papers, 1889–1954, North Dakota Institute for Regional Studies, North Dakota State University, Fargo; and Bishop John Shanley, Diary, 1903, Diocesan Archives of Fargo, courtesy of the Rev. T. William Coyle, C.SS.R., Chancellor, Fargo, N. D. For Bishop Shanley's career, see *St. Mary's Cathedral, 1899–1949* (1949).

62. *The Revised Codes of the State of North Dakota, 1899* (Bismarck, 1899), 698. See also Mariellen MacDonald Noudeck, "Morality Legislation in Early North Dakota, 1889–1914," M.A. thesis, University of North Dakota, Grand Forks, 1964.

63. John Lee Coulter, "Marriage and Divorce in North Dakota," *American Journal of Sociology,* 12 (1906–07):398–416; and Nicolai Rolfsrud Erling, *The Story of North Dakota* (Alexandria, Minn., 1963), 232. See also George B. Winship, "Political History of the Red River Valley," in *History of the Red River Valley* (Grand Forks, 1909), I, 459; "Fargo Was World's Quick Divorce Center for Two Decades," *The Fargo Forum,* Diamond Jubilee Edition, June 4, 1950; William L. O'Neill, *Divorce in the Progressive Era* (New Haven, 1967), 235; and Fargo-Moorhead Centennial Corporation, *A Century Together: A History of Fargo, North Dakota and Moorhead, Minnesota* (1975), 41.

64. The historian cited is O'Neill, *Divorce in the Progressive Era,* 235. Government figures are in Plateris, *100 Years of Marriage and Divorce Statistics,* 34. Cass County figures are from *Register of Actions,* Territory of Dakota, County of Cass, May 1882 to February 1885; *Register of Actions,* District Court, Cass County, N. D., June 1892 to Oct. 1893 and March 1901 to April 1903. The Cass County statistics confirm that more women than men obtained divorces, but because the figures do not indicate migratory decrees it is impossible to determine whether more women than men also secured migratory decrees.

65. *St. Paul Dispatch* is quoted in O'Neill, *Divorce in the Progressive Era,* 235; and the van Vleck case was reported in *The Fargo Forum and Daily Republican,* Jan. 7, 1899.

66. *General Statutes of Oklahoma, 1908* (Kansas City, 1908), 52; and Will T. Little, L. G. Pittman, and R. J. Barker, comps., *The Statutes of Oklahoma, 1890* (Guthrie, 1891) 676–79, 903–5.

67. Ibid., 903–5.

68. *The Statutes of Oklahoma, 1893* (Guthrie, 1893), 875, 877.

69. Frank Dale (Chief Justice), *Reports of Cases Argued and Determined in the Supreme Court of the Territory of Oklahoma,* (Guthrie, 1896), III, 186–

204; *The Daily Leader* [*Guthrie*], Aug. 12, 1892; and *The Daily Times Journal* [*Oklahoma City*], Dec. 26, 1894.

70. Frank Dale (Chief Justice), *Reports of Cases Argued and Determined in the Supreme Court of the Territory of Oklahoma* (Guthrie, 1896), II, 180–228.

71. *The Kingfisher* [*Okla.*] *Free Press*, Sept. 20, 1894; *The Daily Leader* [*Guthrie*], Sept. 9, 1894; *Journal of the Council Proceedings of the Third Legislative Assembly* (1895), 515, 891; and *Territory of Oklahoma Session Laws of 1895* (1895), 107.

72. *Edmond* [*Okla.*] *Sun Democrat,* June 28, 1895; and *Alva* [*Okla.*] *Republican,* May 20, 1896.

73. Quoted in Daniel F. Littlefield, Jr., and Lonnie E. Underhill, "Divorce Seeker's Paradise: Oklahoma Territory, 1890–1897," *Arizona and the West* 17 (1975):23. For an example of a notice by publication see *The [Oklahoma City] Oklahoma Times Journal,* July 15, 1895.

74. Abraham J. Seay, "Governor's Message to Second Legislative Assembly of the Territory of Oklahoma delivered Jan. 19, 1893," *Oklahoma Territory Governors Messages and Reports, 1893–95, 1901–05* (no place, no date), 8; *The Kingfisher* [*Okla.*] *Free Press,* Sept. 6, 1894; *The Daily Ardmoreite* [*Ardmore, Okla.*], Sept. 13, 1894; and *Stillwater* [*Okla.*] *Gazette,* Nov. 19, 1896.

75. *The Daily Leader* [*Guthrie*], March 28, 1894; and *The Oklahoma* [*Oklahoma City*] *Time Journal,* March 30, 1894.

76. *El Reno* [*Okla.*] *News,* Dec. 23, 1898.

77. Ibid., May 29 and Aug. 7, 1896; and *The Daily Oklahoman* [*Oklahoma City*], Jan. 25, 1908. For accounts of the congressional action see *The Indian Chieftain* [*Vinita*], May 21, 1896; and *The Hennessy* [*Okla.*] *Clipper,* June 4, 1896. For revisions of divorce statutes see *Revised Laws of Oklahoma 1910* (St. Paul, Minn., 1912), 1330–37.

78. Quoted in Littlefield and Underhill, "Divorce Seeker's Paradise," 21. Government statistics are found in Plateris, *100 Years of Marriage and Divorce Statistics,* 35.

79. Selected Territorial Divorce Cases, 1890–99, in Territorial Records, Logan County. In the other sampled cases, two petitioners claimed six months, one nine months, twenty-four one year, thirteen two years, one three years, three four years, six five to six years, and one eight years. Two of the sixty-six were undecipherable.

80. Cases #3, *Keller v. Keller,* filed June 17, 1890; 510, *Hawk v. Hawk,* March 2, 1891; 798, *Zeller v. Zeller,* Oct. 18, 1891; 725, *Wilder v. Wilder,* May 10, 1892; 1926, *Davidson v. Davidson,* May 8, 1895; 1949, *Bair v. Bair,* May 28, 1895; 1978, *Thorpe v. Thorpe,* July 25, 1895; 1982, *Mitchell v. Mitchell,* July 26, 1895; and 1958, *Winston v. Winston,* Aug. 1895. All in Territorial Records, Logan County.

Among the cases not cited, *Hawk v. Hawk* was filed on March 2, 1891. Ida married Frank in 1888 in Kansas, where he abandoned her the following year and lived with another woman. He was convicted of bigamy and sentenced to the Kansas State Penitentiary. On April 24, 1891, the Logan County court granted her a divorce, custody of the child, and restoration of her family name. This too may have been a migratory divorce or Ida Hawk may have permanently relocated in Oklahoma Territory.

*Wilder v. Wilder,* filed on May 10, 1892, involved a couple who married in Guthrie in 1891, but perhaps lived outside the Territory, for when Vera filed her divorce petition she claimed the minimum residency of 90 days. When Horace failed to appear she was granted the divorce on June 1. The Wilders appear to have been previous residents of the Territory; it's possible that Vera returned home for her divorce.

*Thorpe v. Thorpe* was filed on July 25, 1895. This couple married in Michigan in 1885 and later moved to Kansas, where Abbie left him in 1894. J. Dayton Thorpe moved to Guthrie with another woman. Abbie followed him to Oklahoma and initiated a divorce action that included his graphic accounts of her verbal and physical abuse of him. Unaccountably, the action was never completed. Because J. Dayton Thorpe was a resident of Oklahoma, the divorce was not migratory.

81. Guest Register, Hotel Springer, Oct. 6, 1892, through Nov. 17, 1893, Oklahoma Territorial Museum, Guthrie.

82. This view is supported by U.S. Dept. of Commerce and Labor, *Marriage and Divorce,* I, 15, 34.

83. Lou V. Chapin, "Uniform Marriage and Divorce Laws," *Good Form* 3 (June 1892):50.

**Chapter 5**

1. *New York Tribune,* July 28, 1879. Figures that support the validity of these fears are found in Alexander A. Plateris, *100 Years of Marriage and Divorce Statistics: United States, 1867–1967* (Rockville, Md., 1973), 28–29; and U.S. Dept. of Commerce and Labor, *Marriage and Divorce, 1867–1906,* 2 vols. (Westport, Conn., 1978 reprint ed.), II, 710, 716, 728.

2. *Report of the National Divorce Reform League, 1887* (Montpelier, Vt., 1887), 15–16. See also Samuel W. Dike, "The National Divorce Reform League," *Our Day: A Record and Review of Current Reform* I (Jan. 1888):49–54.

3. *Report of the National Divorce Reform League, 1886* (Montpelier, Vt., 1887), 4–5.

4. Ibid., 6–7.

5. Ibid., 5–9.

6. Carroll D. Wright, ed., *A Report of Marriage and Divorce in the United States, 1867–1886* (Washington, 1891 rev ed.), 13–15.

7. U.S. Dept. of Commerce and Labor, *Marriage and Divorce, 1867–1906,* I, 3.

8. Wright, ed., *A Report on Marriage and Divorce,* 150–57, 197.

9. Ibid., 197.

10. When number of divorces was compared with population, fifteen states exceeded Iowa in ratio of divorce granted per 100,000 people. U. S. Dept. of Commerce and Labor, *Marriage and Divorce,* I, 62, 64, 72.

11. *Report of the National Divorce Reform League, 1889* (Boston, 1890), 13–15, 20.

12. Max Rheinstein, *Marriage Stability, Divorce, and the Law* (Chicago, 1972), 46.

13. *Report of the National Divorce Reform League, 1891* (Boston, 1892), 8–9.

14. E.D.E.N. Southworth, *Self-Raised; or, From the Depths* (Philadelphia, 1876) and *The Unloved Wife* (New York, 1881). In Henry James' well-known 1880 novel, *Portrait of a Lady,* Isabel Archer also clung to marriage, but for more crass reasons—she wanted to avoid broadcasting "her mistake." See Henry James, *The Portrait of a Lady* (New York, 1975 ed.).

15. Margaret Lee, *Divorce; or, Faithful and Unfaithful* (New York, 1889), 10, 14, 109–11, 218–19, 377, 380–81, 391, 399, 407–8, 411.

16. William Dean Howells, *A Modern Instance* (Bloomington, Ind., 1977 ed.); and *Church Review* 48 (July/Dec. 1886):392. The plot of *Modern Instance* is discussed more fully in Chapter 3 above.

17. Madeleine Vinton Dahlgren, *Divorced. A Novel* (Chicago, 1887), 5–6, 56–67, 205, 211.

18. Margaret Deland, *Philip and His Wife* (Boston, 1894), 73, 401–18, 434. For additional discussion of divorce novels during these years see James H. Barnett, *Divorce and the American Divorce Novel* (New York, 1968), 69–92.

19. Nathan Allen, "Divorces in New England," *North American Review* 130 (June 1880):558–59, 563; Noah Davis, "Marriage and Divorce," *North American Review* 139 (July 1884):30–41; and "Divorce in South Dakota," *The Nation* 56 (Jan. 26, 1893):60.

20. "Children's Side of Divorce," *The Outlook* 70 (Feb. 22, 1902):478–80.

21. *New York Times,* May 25, 1904; and *New York Tribune,* Jan. 27, 1905.

22. Quoted in U. S. Dept. of Commerce and Labor, *Marriage and Divorce,* I, 4.

23. Ibid.

24. Frances E. Willard, letter to "Dearest Susan," Jan. 26, 1898, Ida Husted Harper Collection, Huntington Library, San Marino.

25. Elizabeth Cady Stanton, Susan B. Anthony, and Matilda J. Gage, *History of Woman Suffrage* (New York, 1881, 1882, 1886), I, 722. See also Elisabeth Griffith, "Elizabeth Cady Stanton on Marriage and Divorce: Feminist Theory and Domestic Experience," 233–51, in Mary Kelley, ed., *Woman's Being, Woman's Place: Female Identity and Vocation in American History* (Boston, 1979); Elizabeth Clark, "Matrimonial Bonds: Slavery, Contract and the Law of Divorce in Nineteenth-Century America," Legal History Program Working Papers, Series 1, No. 10, June 1987, Institute for Legal Studies, University of Wisconsin–Madison Law School, 1–37; and William Leach, *True Love and Perfect Union: The Feminist Reform of Sex and Society* (New York, 1980).

26. Clara B. Colby, address titled "Woman in Marriage," No month or day, 1889, in Scrapbook of *Woman's Tribune Clippings, 1883–1891,* Huntington Library, San Marino.

27. Alice Locke Park, letters to "Dear Sir," Nov. 24, 1905, Alice Locke Park Correspondence, Susan B. Anthony Memorial Collection, Huntington Library, San Marino.

28. Replies to Alice Locke Park's letter of Nov. 24, 1905, regarding the Congress on Uniform Divorce Law and printed list of delegates attending the Congress held at the New Willard Hotel, Washington, D.C., Feb. 19–26, 1906, both in Alice Locke Park Correspondence, Susan B. Anthony Memorial Collection, Huntington Library, San Marino.

29. *Proceedings of the Adjourned Meeting of the National Congress on Uniform Divorce Laws, Held at Washington, D.C., February 19, 1906* (Harrisburg, Pa., 1906), 4–8, 20–21, 57–58, 71–72, 126, 139.

30. William L. O'Neill, *Divorce in the Progressive Era* (New Haven, 1967), 254–73.

31. *The Nation* 82 (June 21, 1906):505.

32. The deliberations of the second Congress are found in *Proceedings of the Adjourned Meeting of the National Congress on Uniform Divorce Laws, Held at Philadelphia, Pa., November 13, 1906* (Harrisburg, Pa., 1907). The *New York Tribune* made its dire prophecy on Nov. 16, 1906.

33. Madeleine B. Stern, transcriber, "Two Unpublished Letters from Belva Lockwood," *Signs* I (Autumn 1975):275. Although women could not vote, nothing prohibited them from running for public office.

34. Stanton's statement on slavery is found in Elizabeth Cady Stanton, "The Need of Liberal Divorce Laws," *American Review* 139 (Sept. 1884):243. Her other views are found in Elizabeth Cady Stanton, "Divorce Versus Domestic Warfare," in David J. Rothman and Sheila M. Rothman, eds., *Divorce: The First Debates* (New York, 1987), 560–61.

35. Frances E. Willard, letter to "My dear Friend," [Ida Husted Harper], Jan. 4, 1898, Ida Husted Harper Collection; and Lydia Kingmill Comander, letter to "My dear Mrs. Park," June 2, 1908, Alice Locke Park Correspondence, Susan B. Anthony Memorial Collection, both at the Huntington Library, San Marino.

36. U.S. Dept. of Commerce and Labor, *Marriage and Divorce*, I, 5–6.

37. Ibid., I, 13. That the years between the Civil War and World War I were characterized by changes in other traditional attitudes in addition to divorce is argued in Daniel Scott Smith, "The Dating of the American Sexual Revolution: Evidence and Interpretation," 426–38, in Michael Gordon, ed., *The American Family in Social-Historical Perspective* (New York, 1978).

38. C. LaRue Munson, "The Divorce Question in the United States," *Yale Law Journal* 18 (April 1909):387–98.

39. Jane Burr, *Letters of a Dakota Divorcee* (Boston, 1909), 9–11, 21, 54, 65, 74, 89–90.

40. *New York Tribune*, Aug. 12 and 15, 1911; and James H. Hawley, "Uniformity of Marriage and Divorce Laws," 160–64, in Julia E. Johnson, comp., *Selected Articles on Marriage and Divorce* (New York, 1925).

41. Quoted in Nelson M. Blake, *The Road to Reno: A History of Divorce in the United States* (New York, 1962), 146–47.

42. Charles F. Thwing and Carrie F. Butler Thwing, *The Family: An Historical and Social Study* (Boston, 1913), 198, 219.

43. Mary A. Livermore, "Women's Views of Divorce," *North American Review* 150 (Jan. 1890):115–16.

44. Ibid.; and Marguerite O. B. Wilkinson, "Education as a Preventive

of Divorce," 88–97, in Johnson, *Selected Articles*. For further discussion of sexuality during these years see Carl N. Degler, "What Ought To Be and What Was: Women's Sexuality in the Nineteenth Century," 403–25, in Gordon, ed., *The American Family*; John S. Haller, Jr., "From Maidenhood to Menopause: Sex Education for Women in Victorian America," 71–85, in Thomas L. Altherr, ed., *Procreation or Pleasure?: Sexual Attitudes in American History* (Malabar, Fla., 1983); and Peter Gay, *The Bourgeois Experience: Victoria to Freud* (New York, 1986), Vol. II, *The Tender Passion*.

45. Livermore, "Women's Views of Divorce," 116.

46. For a discussion of these and other free divorce advocates see O'Neill, *Divorce in the Progressive Era*, 89–125.

47. Walter F. Willcox, *The Divorce Problem: A Study in Statistics* (New York, 1891), 61, 66–70, 72–73.

48. George E. Howard, *A History of Matrimonial Institutions*, 3 vols. (Chicago, 1904), III, 160–223.

49. William O'Neill, "Divorce in the Progressive Era," 146–47, in Michael Gordon, ed., *The American Family* (New York, 1978).

50. Edward Alsworth Ross, "Significance of Increasing Divorce," 54–62, in Johnson, comp., *Selected Articles on Marriage and Divorce*.

51. James P. Lichtenberger, *Divorce: A Study in Social Causation* (New York, 1909), 12–13, 52–63, 142–50, 219–25.

52. "Divorces in Massachusetts, 1860–1904," *Massachusetts Labor Bulletin* 40 (Dec. 1906):447–61.

53. U.S. Dept. of Commerce and Labor, *Marriage and Divorce*, I, 11–13, 22–24.

54. Ibid., 25–26, 28.

55. Ibid., 45–46. A female newspaper reporter who divorced is described in Walter E. Kaloupek, "Alice Nelson Page: Pioneer Career Woman," *North Dakota Historical Quarterly* 13 (1946):71–79; and a divorced doctor is described in Estelle C. Laughlin, "Dr. Georgia Arbuckle Fix: Pioneer," 21–39, in Asa B. Wood, ed., *Pioneer Tales of the North Platte Valley* (Gering, Neb., 1938).

56. Norton B. Stern, "Denouement in San Diego in 1888," *Western States Jewish Historical Quarterly* 11 (1978):49–55. For a discussion of the weakening Jewish tradition in California, see William M. Kramer and Norton B. Stern, "An Issue of Jewish Marriage and Divorce in Early San Francisco," *Western States Jewish History* 21 (Oct. 1988):46–57. An example of a German couple on the brink of divorce is found in *Mahan v. Schroeder*, 1907, transcript at the McLean County Historical Society, Bloomington, Ill.

57. U.S. Dept. of Commerce and Labor, *Marriage and Divorce*, I, 277–78, 287–88, 295–96.

58. Ibid., 297–98, 302, 323; and *Gibson v. Gibson*, granted Feb. 11, 1881, Albermarle County, Va., Chancery Orders, XII, 1880–82, pp. 199–200, University of Virginia Library, Charlottesville.

59. U.S. Dept. of Commerce and Labor, *Marriage and Divorce*, I, 310–11.

60. Ibid., I, 20–22. A recent study has shown that at least in Boston, African-American families were more stable than whites usually thought them to be. See Elizabeth H. Pleck, "The Two-Parent Household: Black

Family Structure in Late Nineteenth-Century Boston," *Journal of Social History* 6 (Fall 1972):3–31.

61. Register, Mary McLeod Bethune Papers, Armistad Research Center, Tulane University, New Orleans; and "Mary McLeod Bethune," 77, in Barbara Sicherman and Carol Hurd Green, eds., *Notable American Women: The Modern Period* (Cambridge, Massachusetts, 1980).

62. Hiram S. Pomeroy, *The Ethics of Marriage* (New York, 1888), 55–56, 172.

63. Eliza Chester, *The Unmarried Woman* (New York, 1892), 17, 20, 71–72.

64. B.G. Jefferis, M.D., *Search Lights on Health; or, Light on Dark Corners. How to Love, How to Court, How to Marry and How to Live* (Naperville, Ill., 1894), 142–49, 150–55, 165.

65. Charles E. Sargent, *Our Home; or, Influences Emanating from the Hearthstone* (Springfield, Mass., 1899), 266–72.

66. Mary R. Melendy, *Perfect Womanhood* (1903), 39–40, 42, 47–48, 50–53.

67. Anna B. Rogers, *Why American Marriages Fail* (Boston, 1909), 6–7, 16.

68. A Member of the New York Bar, *How to Get a Divorce* (New York, 1859).

69. Lou V. Chapin, "Uniform Marriage and Divorce Laws," *Good Form* 3 (June 1892):50.

**Chapter 6**

1. Sara Bard Field, Diary, 1913, Wood Collection, Huntington Library, San Marino.

2. That the new morality appeared well before the U.S. entry into World War I is argued in James R. McGovern, "The American Woman's Pre-World War I Freedom in Manner and Morals," *Journal of American History* 55 (Sept. 1968):315–33. For further discussion of Progressivism and divorce, see William L. O'Neill, *Divorce in the Progressive Era* (New Haven, 1967), esp. 254–73.

3. "The Most Difficult Problem of Modern Civilization," *Current Literature* 18 (1910):59.

4. William Brevda, "Love's Coming-of-Age: The Upton Sinclair–Harry Kemp Divorce Scandal," *North Dakota Quarterly* 51 (1983):59–77. The court believed that Upton Sinclair was guilty of collusion in the adultery of his wife by condoning her affair with Kemp. Collusion between husband and wife was a legal reason to deny divorces in New York. For a discussion of the effect of media presentations of divorce on public attitudes see John D. Stevens, "Social Utility of Sensational News: Murder and Divorce in the 1920s," *Journalism Quarterly* 62 (1985):53–58.

5. Emma Goldman, "The Tragedy of Women's Emancipation," 506–16, in Alice S. Rossi, ed., *The Feminist Papers* (New York, 1973). See also

Alix Kates Shulman, ed., *Red Emma Speaks: Selected Writings and Speeches by Emma Goldman* (New York, 1972).

6. Sara Bard Field, "Notes concerning the letter of Charles Erskine Scott Wood to his son Erskine, written July 22, 1927," 1938 and "Notes regarding her divorce," ca. 1960, Wood Collection, Huntington Library, San Marino. Another California woman of the period who thought that divorce was indicated when love evaporated was Anita M. Baldwin, who divorced her second husband in 1915. Anita M. Baldwin, Fragment of a Note Concerning her Husband, 1915 and Divorce Decree, July 1, 1915, Baldwin Collection, Huntington Library, San Marino.

7. Field, Diary and Notes regarding her divorce, Wood Collection, Huntington Library.

8. Albert Ehrgott, letter to "Dear Mr. Wood," Jan. 3, 1913, and Albert Ehrgott, letter to "My dear Mary," May 1, 1913, all in Wood Collection, Huntington Library.

9. Albert Field Erhgott to "My dear Pops (Wood)," July 29, 1918, Wood Collection, Huntington Library.

10. Albert Erhgott, letter to "Dear Sir (Wood)," June 25, 1918, and to "Dear Mrs. Sara Bard Field," Dec. 6, 1918, Wood Collection, Huntington Library.

11. Charles Erskine Scott Wood, letter to "My dear Son Erskine," July 22, 1927, Wood Collection, Huntington Library. For a discussion of broadening attitudes and a decline in stigma toward divorce see Blake, *Road to Reno*, 226–29.

12. Ibid.

13. Alfred Cahen, *Statistical Analysis of American Divorce* (New York, 1932), 21; and Hugh Carter and Paul C. Glick, *Marriage and Divorce: A Social and Economic Study* (Cambridge, Mass., 1970), 17, 38, 57.

14. For discussions of rural divorce rates see George W. Hill and James D. Tarver, "Marriage and Divorce Trends in Wisconsin, 1915–1945," *Milbank Memorial Fund Quarterly* 30 (Jan. 1952), 5–17; and N. Ruth Wood, "Missouri Marriage and Divorce Statistics, 1940–47," *Journal of the Missouri Bar* 4 (March 1948):36–39. Other trends and statistics for 1928 are found in Cahen, *Statistical Analysis of American Divorce*, 22–38. The effects of wartime mobilization and high unemployment are analyzed in Samuel H. Preston and John McDonald, "The Incidence of Divorce Within Cohorts of American Marriages Contracted Since the Civil War," *Demography* 16 (Feb. 1979):1–25.

15. A discussion of African-American divorce can be found in Carter and Glick, *Marriage and Divorce*, 66–70, 77. Quote is from Tom E. Terrill and Jerrold Hirsch, *Such as Us: Southern Voices of the Thirties* (Chapel Hill, 1978), 269.

16. Cahen, *Statistical Analysis of Divorce*, 16–17; and Carter and Glick, *Marriage and Divorce*, 27, 54. Divorces of bed and board were also still possible in some states, especially in the South. Robert Ruffin Barrows, Jr., for example, sought a divorce of bed and board from his wife of twenty-five years in Louisiana in 1925. Robert R. Barrow Family Papers, 1925, Special Collections, Tulane University, New Orleans.

17. Mrs. Edward Franklin White, "Uniform Marriage and Divorce Laws in the United States," *Current History Magazine, New York Times* 18 (May 1923):246–50.

18. Arthur Capper, "Proposed Amendment to the Constitution," and "Uniform Divorce Bill," 241–45, in Julia Johnson, ed., *Selected Articles on Marriage and Divorce* (New York, 1925).

19. Anonymous, "Question of Divorce," *Freeman* 7 (April 25, 1923):150–51; and Albert C. Ritchie, "Back to States' Rights," *World's Work* 47 (March 1924):525–29.

20. Ralph R. Knapp, "Divorce in Washington," *Washington Historical Quarterly* 5 (April 1914):122–29.

21. For Lord Russell's divorce see Duncan Crow, *The Edwardian Woman* (London, 1978), 174–79.

22. George A. Bartlett, *Men, Women and Conflict* (New York, 1931), 11.

23. Nelson M. Blake, *The Road to Reno: A History of Divorce in the United States* (New York, 1962), 153–54.

24. *Nevada State Journal*, Feb. 1, 1913.

25. Ibid., Feb. 8, 18, 1913.

26. Ibid., Feb. 24, 1915.

27. Ibid., March 19, 1927.

28. Ibid., March 7, 17, 20, 1931.

29. Ibid., March 9, 1931.

30. Paul H. Jacobson and Pauline F. Jacobson, *American Marriage and Divorce* (New York, 1959), 103–4.

31. Interview A, July 27, 1989. held in Richmond, Va. Interviewees' names have been omitted to protect their privacy.

32. Interview B, Oct. 17, 1989, Des Moines, Iowa.

33. Interview C, July 21, 1990, Milwaukee, Wisc.

34. Quoted in Blake, *Road to Reno,* 175.

35. *Haddock v. Haddock* 201 U.S. 562 (1906).

36. *Williams et al. v. North Carolina* 317 U.S. 287 (1942).

37. *Lambert v. Lambert* 41 NYS 2d 840 (1943). In 1944, a specialist in constitutional law suggested that Congress establish the minimum requirements to be met by a state so that its divorces would be recognized by other states. See Carl Mason Franklin, " 'The Dilemma of Migratory Divorces' A Partial Solution Through Federal Legislation," *Oklahoma Law Review* (Aug. 1948):151–70.

38. *Williams et al. v. North Carolina* 325 U.S. 226 (1945). See also Thomas R. Powell, "And Repent at Leisure: The Unhappy Lot of Those Whom Nevada Hath Joined Together and North Carolina Hath Put Asunder," *Harvard Law Review* 58 (Sept. 1945):930–1017; and William G. Ruymann, "The Effect of Nevada Divorce Decrees Out of the State," *Nevada State Bar Journal* 14 (Oct. 1949):239–55. For an analysis of the effect of these Supreme Court rulings on an individual state (Connecticut), see John S. Gilman, "Extraterritorial Divorce," *Connecticut Bar Journal* 23 (Sept. 1949):298–314.

39. *Crouch v. Crouch* 169 P, 2d 897 (1946). For an analysis of rulings in other migratory divorce cases see J. H. C. Morris, "Divisible Divorce," *Har-*

*vard Law Review* 64 (June 1951):1287–1303. For the argument that it might be better to remain in one's home state to get a divorce, see R. Dale Vliet, "A Foreseeable End to Migratory Divorces," *Oklahoma Law Review* 10 (Nov. 1957):432–38.

40. Robert M. Bozeman, "The Supreme Court and Migratory Divorce: A Re-examination of an Old Problem," *American Bar Association Journal* 37 (Feb. 1951):107–10, 168–71. Some experts maintained that the Capper Amendment to the U.S. Constitution that would lead to uniform divorce laws in every state was the solution to these problems. See Delbert L. McLaughlin, "The Migratory Divorce," *Kentucky Law Journal* 38 (May 1950):600–608.

41. Peter Nash Swisher, "Foreign Migratory Divorces: A Reappraisal," *Journal of Family Law* 21 (1982–83):9–52.

42. *Williams et al. v. North Carolina* 317 U.S. 287 (1942). For an example of a recent case in which the home state accepted another state's divorce decree, but reserved jurisdiction over the couple's financial settlement, see *Squitieri v. Squitieri* 481 A.2d 585 (N.J.Super.Ch. 1984). For other discussions of full faith and credit for migratory divorces see "Full Faith and Credit and the Out of State Divorce," *DePaul Law Review* 4 (Autumn–Winter 1954):73–79; Henry J. Foster, Jr., "For Better or Worse? Decisions Since *Haddock v. Haddock*," *American Bar Association Journal* 47 (Oct. 1961):963–67; Harriet E. Miers, "Full Faith and Credit— Procedural Limitation Bars Sister State's Collateral Attack on Jurisdiction," *Southwestern Law Journal* 22 (Oct. 1968):662–75; Karl M. Rodman, "Bases of Divorce Jurisdiction," *Illinois Law Review* 39 (March–April 1945):343–66; George W. Stumberg, "The Migratory Divorce," *Washington Law Review* 33 (Winter 1958):331–42; and P.N. Swisher, "Foreign Migratory Divorces: A Reappraisal," *Journal of Family Law* 21 (Nov. 1982):9–52.

43. Quoted in Susan Smith, "Always at Your Feet" (autobiography), undated, University of Virginia Library Special Collections, Charlottesville.

44. This sample was selected unscientifically. Respondents were pointed out to me by their family members or friends who knew that they had divorced before 1945. The sample included thirty women and fifteen men who were all midwestern in background. Two were Catholics and the rest Protestant or unchurched. The interviews were conducted between Nov. 1983 and July 1989.

45. Lawrence M. Friedman, "Rights of Passage; Divorce Law in Historical Perspective," *Oregon Law Review* 63 (1984):659–60; and Blake, *Road to Reno*, 189–225.

46. For a fuller discussion of collusion see Max Rheinstein, *Marriage Stability, Divorce, and the Law* (Chicago, 1972), 55–105.

47. William E. Carson, *The Marriage Revolt: A Study of Marriage and Divorce* (New York, 1915), 436, 445.

48. Charles G. Norris, *Brass: A Novel of Marriage* (New York, 1921), 105, 157–58, 194–97, 319, 421–23; and Annegret Ogden, "Love and Marriage: Five California Couples," *The Californians* 5 (July/Aug. 1987):15–19. Another writer who linked divorce and individualism was George L.

Koehn, "Is Divorce a Social Menace?" *Current History Magazine, New York Times* 16 (May 1922):297.

49. Robert S. Lynd and Helen Merrell Lynd, *Middletown: A Study in Contemporary American Culture* (New York, 1929); and Elaine Tyler May, *Great Expectations: Marriage and Divorce in Post-Victorian America* (Chicago, 1980). See also Elaine Tyler May, "Marriage and Divorce in Los Angeles, 1890–1920," Ph.D. dissertation, University of California, Los Angeles, 1975; and "In-Laws and Out-Laws: Divorce in New Jersey, 1890–1925," 31–41, in Mary R. Murrin, ed., *Women in New Jersey History* (Trenton, 1985). The possible relationship of heightened consumerism to an increase in divorce is discussed in Elaine Tyler May, "The Pressure to Provide: Class, Consumerism, and Divorce in Urban America, 1880–1920," *Journal of Social History* 12 (1978):180–93.

Changes in American courtship are discussed in Ellen K. Rothman, *Hands and Hearts: The History of Courtship in America* (New York, 1984), esp. Parts Three and Four; and in Beth L. Bailey, *From Front Porch to Back Seat: Courtship in Twentieth Century America* (Baltimore, 1988). A survey of changing images of women is found in Martha Banta, *Imaging American Women: Ideas and Ideals in Cultural History* (New York, 1987); and of women's work in Mary E. Cookingham, "Combining Marriage, Mother-hood, and Jobs Before World War II: Women College Graduates, Classes of 1905–1935," *Journal of Family History* 9 (Summer 1984):178–95. Chang-ing expectations of family life are discussed in Steven Mintz and Susan Kellogg, *Domestic Revolutions: A Social History of American Family Life* (New York, 1987). The effect of the Depression on family life is found in Wini-fred D. Wandersee Bolin, "The Economics of Middle-Income Family Life: Working Women During the Great Depression," *Journal of American History* 65 (June 1978):60–74; and of World War II in Karen Tucker Anderson, *Wartime Women: Sex Roles, Family Relations, and the Status of Women during World War II* (Westport, Conn., 1981), esp. chap. 3, "The Family in War-time," 75–121. Demographic factors affecting the family are discussed in Robert V. Wells, "Demographic Change and the Life Cycle of American Families," *Journal of Interdisciplinary History* 2 (Autumn 1971):276–77. See also Robert V. Wells, *Revolutions in Americans' Lives: A Demographic Perspec-tive on the History of Americans, Their Families, and Their Society* (Westport, Conn., 1982).

50. See for example Arthur Calhoun, *A Social History of the Family Since the Civil War* (Cleveland, 1919), III, 268–81; Koehn, "Is Divorce a Social Menace?" 294–99; and Gustavus Myers, *Current History Magazine, New York Times* 14 (Aug. 1921), 116–21. Sex as a modern problem is discussed in Henry Seidel Canby, "Sex and Marriage in the Nineties," *Harper's Magazine* (Sept. 1934):427–36. Women's emancipation as a causal factor in divorce is discussed in Elaine Tyler May, "In-Laws and Out-Laws." The declining power of stigma and anti-divorce attitudes to keep people in marriages was noted by Lynd and Merrell Lynd, *Middletown in Transition*, 161.

51. Cahen, *Statistical Analysis*, 90–91, 97.

52. Isabel Drummond, *Getting a Divorce* (New York, 1931), 21, 58.

53. Cahen, *Statistical Analysis*, 137.

54. Alexander A. Plateris, *100 Years of Marriage and Divorce Statistics: United States, 1867–1967* (Rockville, Md., 1973), 2–3.

55. State divorce rates are found in Plateris, *100 Years of Marriage and Divorce Statistics*, 34–35. Numbers of Linn County divorces are from *Record of Divorces, 1928–1944, Linn County, Iowa*, Linn County Court House, Cedar Rapids. The statistics were compiled by the author. The 1928–44 period was chosen for what it might reveal about changes in numbers of divorces during the Depression years. The data, however, showed no significant rise or drop in divorces during those years, but it did reveal the difficulties of analyzing divorce cases. For a description of Linn County and its major city, Cedar Rapids, see Ernie Danek, *Tall Corn and High Technology* (Woodland Hill, Calif., 1980).

56. Impotence was the most commonly cited cause for annulment. See for example cases #55321, 40958, and 48869, Linn County Courthouse. Names of litigants have been omitted from all Linn County cases to preserve the privacy of their families.

57. Cases #39068 and 50739, Linn County Courthouse. Such variation in documents and resulting interpretations often led to debates about whether divorce law helped or hurt women. For such a debate see Claude D. Stout, "The Legal Status of Women in Wisconsin," *Marquette Law Review* 14 (Feb./June 1930):155–65; and Julia B. Dolan, "Another Version of the Legal Status of Women in Wisconsin," *Marquette Law Review* 15 (April 1931):139–57.

58. Cases #39068, 37932, 41557, 42003, 39124, Linn County Courthouse.

59. Case #42022, Linn County Courthouse.

60. Case #55585, Linn County Courthouse.

61. Case #52678, Linn County Courthouse.

62. Isabel Drummond, *Getting a Divorce* (New York, 1931), 93–96. For an overview of changing divorce law see Lawrence M. Friedman, "Rights of Passage: Divorce Law in Historical Perspective," *Oregon Law Review* 63 (1984):649–69; Rheinstein, *Marriage Stability, Divorce, and the Law;* and Roderick Phillips, *Putting Asunder: A History of Divorce in Western Society* (Cambridge, Eng., 1988).

63. The sample of 900 cases was derived from *Register of Divorces, 1928–1944*, Linn County Courthouse. The data was compiled by the author.

64. Numerous cruelty decisions are reviewed in *Supplement to the Code of Iowa, 1913* (Des Moines, 1914), 1366–67. See also Gallaher, *Legal and Political Status of Women*, 67–68. A similar trend occurred in other parts of the country as well. See for example Jane Turner Censer, " 'Smiling Through Her Tears'; Ante-Bellum Southern Women and Divorce," *American Journal of Legal History* 25 (Jan. 1981):24–47.

65. Case #38441, Linn County Courthouse.

66. Case #52681, Linn County Courthouse.

67. Case #53723, Linn County Courthouse.

68. Case #42003, Linn County Courthouse. In most felony cases, the crime went unnamed. See for example case #57028.

69. Case #50739 and 52631, Linn County Courthouse.

70. Case #41557, Linn County Courthouse.
71. Case #46669, Linn County Courthouse. For another example of a husband charging his wife with adultery, see case #55725.
72. Case #37932, Linn County Courthouse.
73. Case #39124, Linn County Courthouse.
74. Cases #39792 and 42022, Linn County Courthouse.
75. Ibid.
76. Ibid.
77. Ibid.
78. Anonymous, "Thrice Married," *Harper's Monthly* 176 (April 1938):545–53.
79. Cases #45509 and 52353, Linn County Courthouse.
80. These data were also derived from the sample of 900 cases, *Record of Divorces, 1928–1944*, Linn County Courthouse.
81. Case #56132, Linn County Courthouse.
82. Case #44499, Linn County Courthouse.
83. Case #55257, Linn County Courthouse.
84. Amelie Rives, *Shadows of Flames: A Novel* (New York, 1914); and Helen Lojek, "The Southern Lady Gets a Divorce: 'Saner Feminism' in the Novels of Amelie Rives," *Southern Literary Journal* 12 (1979):47–69; Mari Sandoz, *Old Jules* (Lincoln, Neb., 1935); and Melody Graulich, "Every Husband's Right: Sex Roles in Mari Sandoz's *Old Jules*," *Western American Literature* 18 (April 1983):3–20.
85. Field, "Notes concerning the letter of Charles Erskine Scott Wood," Wood Collection, Huntington Library, San Marino.

**Chapter 7**

1. Figures for the late 1980s are from Claire Safran, "Partnership and Prejudice: The Economics of Divorce," *Lear's* 2 (May 1989):32. Mister Rogers was quoted in "Overheard," *Newsweek*, special edition on "The 21st Century Family" (Winter/Spring 1990), 11. The topic of divorce also permeated the media. An example of an anti-divorce versus pro-divorce debate was a "Crossfire" segment on divorce moderated by Patrick Buchanan and Michael Kinsly, CNN television network, 5:30 EST, Jan. 11, 1989. For examples of literary treatments of divorce, see Emily Prager, *Clea & Zeus Divorce* (New York, 1987); Anne Tyler, *Dinner at the Homesick Restaurant* (New York, 1982); and John Updike, *Trust Me* (New York, 1987). Unlike earlier novels, these books do not support or protest divorce nor do they explore causes for divorce. Rather, they tend to take divorce for granted and analyze the marriage that led to divorce.
2. J. Nelson Frierson, "Divorce in South Carolina," *North Carolina Law Review* 9 (1931):275–82. For a discussion of changing attitudes toward women, family, and divorce in South Carolina see Dennis Lee Frobish, "The Family and Ideology: Cultural Constraints on Women, 1940–60," Ph.D. dissertation, University of North Carolina, Chapel Hill, 1983; and

Susan Freda Hult, "Divorce and the Decline of the Patriarchy in South Carolina," M. A. thesis, Clemson University, 1985. For a popular essay claiming that most South Carolinians favored divorce, see Leroy M. Want with W. D. Workman, Jr., "Divorce—A South Carolina Problem," *South Carolina Magazine* 12 (March 1949):33–34.

3. J. D. Sumner, Jr., "The South Carolina Divorce Act of 1949," *South Carolina Law Quarterly* 3 (1951):253–59.

4. An example of a call for reform is Richard H. Wels, "New York: The Poor Man's Reno," *Cornell Law Quarterly* 35 (Winter 1950):303–26. Divorce reform is described in Henry H. Foster, Jr., and Doris Jonas Freed, *The Divorce Reform Law* (Rochester, N. Y., 1970), 1–23.

5. Ibid., 7–8.

6. Quote is from Paul H. Jacobsen, *American Marriage and Divorce* (New York, 1959), 9. See also Hugh Carter and Paui C. Glick, *Marriage and Divorce: A Social and Economic Study* (Cambridge, Mass., 1970), 384–85.

7. Samuel C. Newman, "Trends in Vital Statistics of Marriages and Divorces in the United States," *Marriage and Family Living* 12 (Summer 1950), 89; and John F. Crosby, "A Critique of Divorce Statistics and Their Interpretation," *Family Relations* 29 (Jan. 1980):51–58.

8. "The 'Big D' Means Divorce," *Des Moines Register* April 25, 1986.

9. Robert V. Wells, "Demographic Change and the Life Cycle of American Families," *Journal of Interdisciplinary History* 2 (Autumn 1971):276–77. See also Robert V. Wells, *Revolutions in Americans' Lives: A Demographic Perspective on the History of Americans, Their Families, and Their Society* (Westport, Conn., 1982).

10. Carter and Glick, *Marriage and Divorce*, 82–85, 388–89; Calvin L. Beale, "Increased Divorce Rates Among Separated Persons as a Factor in Divorce Since 1940," *Social Forces* 29 (Oct. 1950):72–74; and Thomas P. Monahan, "The Changing Nature and Instability of Remarriage," *Eugenics Quarterly* 5 (1958):73–85.

11. Carter and Glick, *Marriage and Divorce*, 299; and John Sirjamaki, *The American Family in the Twentieth Century* (Cambridge, Mass., 1953), 165–66. See also Paul H. Jacobson, "Marital Dissolutions in New York State in Relation to Their Trend in the United States," *Milbank Memorial Fund Quarterly* 28 (Jan. 1950):25–42; Thomas P. Monahan, "Desertion and Divorce in Philadelphia," *American Sociological Review* 17 (Dec. 1952):719–27; and Jessie Bernard, *Marriage and Family Among Negroes* (Englewood Cliffs, N.J., 1966), 101–2.

12. Alexander A. Plateris, *Divorce Statistics Analysis: United States, 1964 and 1965* (Rockville, Md., 1969), Vol. I:*Increases in Divorces: United States, 1967* (Rockville, Md., 1970), 1–3; and *Divorces: Analysis of Changes, United States, 1969* (Rockville, Md., 1973), 1–2; and Bureau of the U. S. Census, *Statistical Abstract of the United States, 1985* (Washington, D.C., 1984), 54, 80.

13. Carter and Glick, *Marriage and Divorce*, 58–59, 66–71. The difference between urban and rural rates is discussed in A. R. Mangus, "Marriage and Divorce in Ohio," *Rural Sociology* 14 (June 1949):128–37; Karen Woodrow, Donald W. Hastings, and Edward J. Tu, "Rural-Urban

Patterns of Marriage, Divorce, and Mortality: Tennessee, 1970," *Rural Sociology* 43 (1978):70–86; and Robert R. Reynolds, Jr., et al., "An Exploratory Analysis of County Divorce Rates," *Sociology and Social Research* 69 (1984):109–21.

14. An analysis of dominance among divorced African-American women is found in William M. Chavis and Gladys J. Lyles, "Divorce Among Educated Black Women," *Journal of the National Medical Association* 67 (March 1975):128–34. See also William M. Chavis and Gladys J. Lyles, *Divorce Among Educated Black Women* (Tuskegee Institute, Ala., 1974). Egalitarianism in African-American families is discussed in Charles V. Villie, *Black and White Families: A Study in Complementarity* (New York, 1985). The high divorce among African Americans is discussed in Bernard, *Marriage and Family Among Negroes,* 101–3; James A. Sweet and Larry L. Bumpass, "Differential in Marital Instability of the Black Population: 1970," *Phylon* 35 (1974):323–31; and Robert L. Hampton, "Husband's Characteristics and Marital Disruption in Black Families," *The Sociological Quarterly* 20 (Spring 1979):255–66.

15. Peter Uhlenberg, "Marital Instability among Mexican Americans: Following the Patterns of Blacks?" *Social Problems* 20 (1972):49–56; Isaac W. Eberstein and W. Parker Frisbie, "Differences in Marital Instability among Mexican Americans, Blacks, and Anglos: 1960 and 1970," *Social Problems* 23 (1976):609–21; W. Parker Frisbie, Frank D. Bean, and Isaac W. Eberstein, "Recent Changes in Marital Instability among Mexican Americans: Convergence with Black and Anglo Trends?" *Social Forces* 58 (1980):1205–20; and Richard Griswold Del Castillo, *La Familia: Chicano Families in the Urban Southwest, 1848 to the Present* (Notre Dame, 1984), esp chap. 8, "The Contemporary Chicano Family," 112–26.

16. Carter and Glick, *Marriage and Divorce* 393–94; Howard M. Bahr, "Religious Intermarriage and Divorce in Utah and the Mountain States," *Journal for the Scientific Study of Religion* 20 (1981):351–61; and Leo Driedger, Michael Yoder, and Peter Sawatzky, "Divorce Among Mennonites: Evidence of Family Breakdown," *Mennonite Quarterly Review* 59 (1985):367–82.

17. Carter and Glick, *Marriage and Divorce,* 391–94; and C.K. Cheng and Douglas S. Yamamura, "Interracial Marriage and Divorce in Hawaii," *Social Forces* 36 (Oct. 1957):77–84.

18. Paul H. Jacobson, "Differential in Divorce by Duration of Marriage and Size of Family," *American Sociological Review* 15 (April 1950):235–44; and William Sander, "Women, Work, and Divorce," *American Economic Review* 75 (1985):519–23. See also Carter and Glick, *Marriage and Divorce,* 38–39, 55, 402–3.

19. Several classic studies arguing for the efficacy of divorce upon children were J. Louise Despert, *Children of Divorce* (New York, 1953); F. Ivan Nye, "Child Adjustment in Broken and Unhappy Unbroken Homes," *Journal of Marriage and the Family* 19 (1957):356–61; and Lee G. Burchinal, "Characteristics of Adolescents from Unbroken, Broken, and Reconstituted Families," *Journal of Marriage and the Family* 26 (1964):44–51. For a review of studies of children of divorce prior to 1975, see Lynne Carol

Halem, *Divorce Reform: Changing Legal and Social Perspectives* (New York, 1980), 176–81.

20. For studies focusing upon parent-child relationships, siblings, and therapy, see Jolene Oppawsky, "Family Dysfunctional Patterns During Divorce—From the View of the Children," 139–51, and Lawrence A. Kurdek, "Siblings' Reactions to Parental Divorce," 203–20, both in *Journal of Divorce* 12, 2/3 (1988–89). See also William F. Hodges, *Interventions for Children of Divorce: Custody, Access, and Psychotherapy* (New York, 1986). For the argument that divorce inflicts long-term harm on children, see Judith S. Wallerstein and Sandra Blakeslee, *Second Chances: Men, Women, and Children a Decade After Divorce* (New York, 1989).

21. Ellen Goodman, "Umbrella Gone, These Young Adults Managing Well," appeared in the *Des Moines Register,* July 24, 1990.

22. For an example of a study assessing the impact of property and support awards on children's adjustment to divorce, see Lenore J. Weitzman, *The Divorce Revolution: The Unexpected Social and Economic Consequences for Women and Children in America* (New York, 1985), esp. 318–21, 352–54. For a study taking race into account, see Christine P. Phillips and Charles A. Asbury, "Relationship of Parental Marital Dissolution and Sex to Selected Mental Health and Self-Concept Indicators in a Sample of Black University Freshmen," *Journal of Divorce* 13, 3 (1990):79–92.

23. William E. Carson, *The Marriage Revolt: A Study of Marriage and Divorce* (New York, 1915), 450; Katharine F. Gerould, "Divorce," *Atlantic Monthly* 132 (Oct. 1923):466; and Ben B. Lindsey and Wainwright Evans, *The Companionate Marriage* (New York, 1927).

24. George Thorman, *Broken Homes* (New York, 1947), 15–20; and Robert B. Deen, Jr., "The Present Status of Connivance as a Defense to Divorce," *Vanderbilt Law Review* 3 (Dec. 1949):107. See Steven Mintz and Susan Kellogg, *Domestic Revolutions: A Social History of American Family Life* (New York, 1987), 178–95; and Elaine Tyler May, *Homeward Bound: American Families in the Cold War Era* (New York, 1988), for an analysis of family life during the mid-twentieth century.

25. Charles H. Leclaire, "Reform—The Law of Divorce," *The George Washington Law Review* 17 (April 1949):390. For a thorough summary of divorce provisions in 1949, see W. Russell Arrington, "Jurisdictional Aspects of Divorce, Annulment, and Separate Maintenance"; Warren W. Kriebel, Jr., "Grounds and Defenses in Divorce, Annulment, and Separate Maintenance"; George E. Harbert, "Property Rights as Affected by Divorce, Annulment, and Separate Maintenance"; Harold W. Holt, "The Conflict of Laws in Divorce"; A. J. Scheineman, "Alimony and Its Enforcement"; J. Nelson Young, "Tax Problems Involved in Divorce"; and Paul W. Alexander, "The Follies of Divorce—a Therapeutic Approach to the Problem"—all in *University of Illinois Law Forum* (Winter 1949), 547–711.

26. George Squire, "The Shift from Adversary to Administrative Divorce," *Boston University Law Review* 33 (April 1953):141–75; Florence Perlow Shientag, "Divorce in the United States: Recent Developments," *Women Lawyers Journal* 43 (Winter 1957):7–9, 20–21; and Mary Somerville Jones, "An Historical Geography of Changing Divorce Law," Ph. D.

dissertation, University of North Carolina, Chapel Hill, 1978, pp. 110, 138. One of the earliest states to accept separation as a ground for divorce was Louisiana. See Zue Vance, "Divorce in Louisiana: Grounds and Defenses," *Tulane Law Review* 24 (June 1950):450–51.

27. See for example, Robert V. Sherwin, *Compatible Divorce* (New York, 1969), 291–98; and Nester C. Kohut, *Positive Divorce Reform for America* (Chicago, 1969), 1–17. See also Nelson M. Blake, *The Road to Reno: A History of Divorce in the United States* (New York, 1962), 237–43.

28. Christopher Lasch, "Divorce and the Family in America," *Atlantic Monthly* 218 (Nov. 1966):57–61; and Margaret Mead, "Double Talk about Divorce," *Redbook* 131, 1 (May 1968):47–48.

29. Henry H. Foster, Jr., "Current Trends in Divorce Law," *Family Law Quarterly* 1 (June 1967):21–40; and Doris Jonas Freed and Henry H. Foster, Jr., "Divorce American Style," *Annals of the American Academy of Political and Social Science* 383 (May 1969):71–87.

30. Howard A. Krom, "California's Divorce Law Reform: An Historical Analysis," *Pacific Law Journal* 1 (Jan. 1970):156–81. For practical advice to divorce-seekers, see Harry Walter Koch, *California Marriage and Divorce Laws* (San Francisco, 1969). For advice to attorneys, see Charles W. Johnson, "The Family Law Act: A Guide to the Practitioner," *Pacific Law Journal* 1 (Jan. 1970):147–55.

31. Doris Jonas Freed and Henry H. Foster, Jr., "Divorce in the Fifty States: An Outline," *Family Law Quarterly* 11 (Fall 1977):297–313; and Robert Raphael, Frederick N. Frank, Joanne Ross Wilder, "Divorce in America: The Erosion of Fault," *Dickinson Law Review* 81 (Summer 1977):719–31. For the effect of no-fault laws on the divorce rate, see Michael Wheeler, *No-Fault Divorce* (Boston, 1974), 30–31, 155–56; Walter D. Johnson, *Marital Dissolution and the Adoption of No-Fault Legislation* (Springfield, Ill., 1975), 22–31; and Dorothy M. Stetson and Gerald C. Wright, Jr., "The Effects of Laws on Divorce in the American States," *Journal of Marriage and the Family* 37 (Aug. 1975):537–47.

32. Donna J. Zenor, "Untying the Knot: The Course and Patterns of Divorce Reform," *Cornell Law Review* 57 (April 1972):649–67.

33. Shana Alexander, *State-by-State Guide to Women's Legal Rights* (Los Angeles, 1975), 99.

34. Jessie Bernard, "No News, but New Ideas," in Paul Bohannan, ed., *Divorce and After* (New York, 1970), 3–25.

35. Arnold O. Ginnow and Milorad Nikolic, *Corpus Juris Secundum: A Contemporary Statement of American Law as Derived from Reported Cases and Legislation* (St. Paul, 1986), 36, 43–44, 59–78.

36. Lawrence M. Friedman and Robert V. Percival, "Who Sues for Divorce? From Fault through Fiction to Freedom," *Journal of Legal Studies* 5 (Jan. 1976):61–82.

37. Kenneth D. Sell, "Divorce Advertising—One Year After *Bates*," *Family Law Quarterly* 12 (Winter 1979):275–83. In international advertisements, attorneys assured divorce-seekers of all nationalities that they could get an American divorce cheaply, rapidly, and without traveling to the United States. In a 1988 edition of the *International Herald Tribune*, classi-

fied advertisements informed divorce-seekers that some divorce attorneys even offered the services of agents in such locations as Amsterdam in the Netherlands and in Guam. These advertisements created the impression that a floating, worldwide divorce market existed. See, for example, *International Herald Tribune,* Aug. 11, 1988.

38. Barbara B. Hirsch, *Divorce: What a Woman Needs to Know* (Chicago, 1973), 72–73.

39. Michael Maier, *Divorce and Annulment in the Fifty States* (New York, 1975), 99.

40. Alexandra Peers, "Differences in Divorce Laws Prompting Some People to Shop for the Best State," *Wall Street Journal,* April 14, 1988, p. 31.

41. Barbara Ehrenreich, "No-Fault, No Fair," *New York Times Book Review,* July 27, 1986.

42. For one answer to this query, see Herbert Jacob, *Silent Revolution: The Transformation of Divorce Law in the United States* (Chicago, 1988).

43. Lenore J. Weitzman, "The Divorce Law Revolution and the Transformation of Legal Marriage," 301–48, in Kingsley Davis and Amyra Grossbard-Schechtman, eds., *Contemporary Marriage: Comparative Perspectives on a Changing Institution* (New York, 1985). Weitzman's data and ideas are discussed more fully in Lenore J. Weitzman, *The Divorce Revolution: The Unexpected Social and Economic Consequences for Women and Children in America* (New York, 1985). See also Linda Kauffman and Benjamin Bycel, "Divorce-American Style," *The Center Magazine,* Nov./Dec. 1980, pp. 4–19; and Terry J. Arendell, "Women and the Economics of Divorce in the Contemporary United States," *Signs* 13 (1987):121–35. For a slightly different view of women's poverty, see Mintz and Kellogg, *Domestic Revolutions,* 216–17.

44. Ibid.

45. The major criteria for selecting interviewees were their having obtained a no-fault divorce and being willing to talk frankly about it. They were also selected to represent a range of age groups, affiliations, social classes, ethnic backgrounds, races, and regions of the country. Names are omitted to protect the privacy of the interviewees.

46. Interview #13, Nov. 13, 1989, Los Angeles.

47. Interview #60, Dec. 29, 1989, St. Louis.

48. Interview #23, Jan. 23, 1989, Los Angeles.

49. Frances E. Kobrin, "The Fall in Household Size and the Rise of the Primary Individual in the United States," 69–81, in Michael Gordon, ed., *The American Family in Social-Historical Perspective* (New York, 1978). See also *Minneapolis–St. Paul Star Tribune,* April 10, 1989, which maintained that increasing numbers of both never-married and divorced people were choosing to remain single.

50. Census Bureau figures were reported in Dennis Meredith, "Dad and the Kids," *Psychology Today* 19 (June, 1985):62–67. For the enlarged experience of fatherhood concept see John Demos, *Past, Present, and Personal* (New York, 1986), 62–63. For women's complaints about child support, see "Women's Lament," *Des Moines Register,* March 10, 1985; and for the problems of single parents see *New York Times,* March 9, 1985.

51. "Parents Without Partners," 649–50, in Donald Scott and Bernard Wishy, eds., *America's Families: A Documentary History* (New York, 1981).

52. Noel Perrin, "Middle-Age Dating," *New York Times Magazine,* July 6, 1986, p. 37; and "Singles in Agriculture," *Waterloo [Iowa] Courier,* Nov. 24, 1986.

53. Census data reported in the *Washington Post,* May 12, 1989. For a profile of co-habitants and reasons for co-habitation, see Graham B. Spanier, "Cohabitation in the 1980s: Recent Changes in the United States," 91–111, in Davis and Grossbard-Schechtman, *Contemporary Marriage.*

54. *Marvin v. Marvin* 557 P.2d 106, 18 Cal.3d 660, 134 Cal.Rptr. 815 (1979).

55. Should Couples Live Together First," *New York Times Magazine,* Oct. 20, 1985; and "Study: Live Together Now, Not Happily Ever After," *The Baltimore Sun,* June 25, 1989.

56. Mintz and Kellogg, *Domestic Revolutions,* 216–26; and Shirley Maxwell Jones, "Divorce and Remarriage: A New Beginning, A New Set of Problems," *Journal of Divorce* 2 (Winter 1978):217–27.

57. Mintz and Kellogg, *Domestic Revolutions,* 226.

58. Lynn K. White and Alan Booth, "The Quality and Stability of Remarriages: The Role of Stepchildren," *American Sociological Review* 50 (Oct. 1985):689–98. See also Robert Garfield, "The Decision to Remarry," *Journal of Divorce* 4 (Fall 1980):1–10.

59. Micaela di Leonardo, "The Female World of Cards and Holidays: Women, Families, and the Work of Kinship," *Signs* 12 (1987):440–53; Paul Bohannan, "Divorce Chains, Households of Remarriage, and Multiple Divorcers," 115, in Paul Bohannan, ed., *Divorce and After* (New York, 1970).

60. Wendy Leigh, "Family Gatherings Can Never Be the Same Again," *New Woman,* Dec. 1985, pp. 103–4.

61. "Dear Abby," *Waterloo [Iowa] Courier,* July 11, 1988.

62. Interviewees were selected because they were involved in, or had recently been involved in, a blended family situation. They represented various age groups, religions, occupations, social classes, ethnic backgrounds, races, and regions of the country. "Bill and Alice," interview with "Alice," Nov. 10, 1988, Norfolk, Va.

63. "Sarah and Phil," interview with "Phil," April 23, 1987, and "Sheila and Keith," interview with "Sheila," April 24, 1987, both in New York City.

64. Barbara Foley Wilson, *Remarriages and Subsequent Divorces: United States* (Hyattsville, Md., 1989), 1–3.

65. Ibid.

66. Advertisement, "Mission of the Bells Wedding Chapel," *What's On in Las Vegas,* Aug. 23–Sept. 5, 1988.

67. "I Do, I Do, I Do," *Newsweek,* Oct. 7, 1985, p. 84.

68. Interview #7, June 3, 1988, Oklahoma City.

69. Leslie Aldridge Westoff, *The Second Time Around: Remarriage in America* (New York, 1977), 60–62; "Premarital Contracts," *Los Angeles Times,* Aug. 19, 1986; Carole Phillips, "Terms of Agreement: Remarriage and

Money," *New Woman,* May 1985, p. 135; and "Prenups Are Up," *New Woman* Nov. 1989, p. 42. Quote is from the article by Phillips.

70. Bernard, *The Future of Marriage,* 225–26; and Bohannan, "Divorce Chains," 123. For a critique of figures on redivorce, see Randal D. Day and Wade C. Mackey, "Redivorce Following Remarriage: A Re-evaluation," *Journal of Divorce* 4 (Spring 1981):39–47.

71. Leonard Slater, "The Disgrace of Hollywood—First Divorce Chart Ever Compiled," *McCall's Magazine* 86 (March 1959):50–52, 141–42, 145, 147. The public's interest in actors' divorces dates back to the pre-Civil War years. See "Charles O'Conor and the Forrest Divorce Case" (1851 case), *New York Law Review* 3 (Jan. 1925):8–12.

72. "I Do . . . I Do!" *USA Weekend,* Sept. 29–Oct. 1, 1989, pp. 4–5.

73. Nathaniel Fishman, *Marriage: This Business of Living Together* (New York, 1946), 246–91.

74. Jacques Bacal and Louise Sloane, *ABC of Divorce* (New York, 1947), 7–20.

75. Benjamin Spock, *The Commonsense Book of Baby and Child Care* (New York, 1946); and Nancy Pottishman Weiss, "Mother, the Invention of Necessity: Dr. Benjamin Spock's *Baby and Child Care,*" *American Quarterly* 24 (Winter 1977):519–46. In 1952, Ashley Montagu's *Natural Superiority of Women* reinforced Spock's message by emphasizing that women's "superiority" derived from their ability to give "mother love," a crucial ingredient in human development and progress. See Ashley Montagu, *The Natural Superiority of Women* (New York, 1953).

76. Special Issue on Women, *Life* 41 (Dec. 24, 1956):41–48.

77. Ibid., 27. For a fuller version of Mead's views concerning women and men see Margaret Mead, *Male & Female* (New York, 1949).

78. Ibid., 73, 109.

79. Regina Markell Morantz, "The Scientist as Sex Crusader: Alfred C. Kinsey and American Culture," 145–66, in Thomas L. Altherr, ed., *Procreation or Pleasure? Sexual Attitudes in American History* (Malabar, Fla., 1983).

80. Cynthia E. Harrison, "A 'New Frontier' for Women: The Public Policy of the Kennedy Administration," *Journal of American History* 67 (Dec. 1980):630–46; Arland Thornton and Deborah Freedman, "Changes in the Sex Role Attitudes of Women, 1962–1977: Evidence from a Panel Study," *American Sociological Review* 144 (Oct. 1979):831–42; and Betty Friedan, *The Feminine Mystique* (New York, 1963). See also Sara M. Evans, *Personal Politics: The Roots of Women's Liberation in the Civil Rights Movement and the New Left* (New York, 1979).

81. Samuel G. Kling, *The Complete Guide to Divorce* (New York, 1963), 28–39, 173–79.

82. Ibid., 147–66; and Eric W. Johnson, *Love and Sex in Plain Language* (Philadelphia, 1965).

83. Morton M. Hunt, *The World of the Formerly Married* (New York, 1966), 4–22.

84. William H. Masters and Virginia E. Johnson, "Settling Sexual Conflicts," *The Reader's Digest,* April 1975, pp. 84–87. See also William H. Masters and Virginia E. Johnson, *The Pleasure Bond* (New York, 1970).

85. Jerry Greenwald, *Creative Intimacy* (New York, 1975), 30; Clare Boothe Luce, "Can Marriage Be Happy?" *The Reader's Digest,* April 1975, pp. 81–84; and Jean Stapleton and Richard Bright, *Equal Marriage* (New York, 1976).

86. Nena O'Neill and George O'Neill, *Open Marriage: A New Life Style for Couples* (New York, 1973); John Powell, *The Secret of Staying in Love* (Niles, Ill., 1974); George R. Bach and Peter Wyden, *The Intimate Enemy* (New York, 1968); George R. Bach and Ronald M. Deutsch, *Pairing: How to Achieve Genuine Intimacy* (New York, 1970); and George R. Bach and Herb Goldberg, *Creative Aggression: The Art of Assertive Living* (New York, 1974). Other examples of how-to-do-it marriage guides are David Viscott, *How to Live With Another Person* (New York, 1974); Jerry Gillies, *My Needs, Your Needs, Our Needs* (New York, 1974); Daniel Goldstine, Katherine Larner, Shirley Zuckerman, and Hilary Goldstine, *The Dance-Away Lover* (New York, 1977); Zev Wanderer and Erika Fabian, *Making Love Work* (New York, 1979); Dan Greenburg and Suzanne O'Malley, *How to Avoid Love and Marriage* (New York, 1983); and Francine Klagsbrun, *Married People: Staying Together in the Age of Divorce* (New York, 1985). No-fault marriage is described in Marcia Lasswell and Norman M. Lobsenz, *No-Fault Marriage* (New York, 1976). For a review of the self-help literature in the late 1970s, see Ellen Ross, "The Love Crisis: Couples Advice Books of the Late 1970s," *Signs* 6 (Autumn 1980):109–22.

87. James T. Friedman, *The Divorce Handbook: Your Basic Guide to Divorce* (New York, 1982).

88. Women in Transition, Inc., *Women in Transition* (New York, 1975), esp. 1–9.

For other examples of women's handbooks see Maxine Schnall, *Your Marriage* (New York, 1976) and Jane Wilkie, *The Divorced Woman's Handbook* (New York, 1980). For a discussion of such books see Barbara Ehrenreich and Deidre English, *For Her Own Good: 150 Years of the Experts' Advice to Women* (New York, 1978).

89. Letter to Glenda Riley from J. W. Wieser, Director, Divorce Centers, Waterloo, Iowa, dated Oct. 27, 1988, in author's possession.

90. Mel Krantzler, *Creative Divorce: A New Opportunity for Personal Growth* (New York, 1973) and Melba Colgrove, Harold H. Bloomfield, Peter McWilliams, *How to Survive the Loss of a Love* (New York, 1976). Other examples of how-to-cope books are Debora Phillips with Robert Judd, *How to Fall Out of Love* (New York, 1978); Bruce Fisher, *Rebuilding When Your Relationship Ends* (San Luis Obispo, Calif., 1981); and Judith Sills, *How to Stop Looking for Someone Perfect & Find Someone to Love* (New York, 1984).

91. Catherine Napolitane with Victoria Pellegrino, *Living & Loving After Divorce* (New York, 1977); Howard M. Halpern, *How to Break Your Addiction to a Person* (New York, 1982); Eric Weber and Steven S. Simring, *How to Win Back the One You Love* (New York, 1983).

92. Shawn D. Lewis, "Divorce: How to Survive One of the Worst Times of Your Life," *Ebony Magazine* 34 (July 1979):53, 55, 58–59; and Lynn Norment, "Divorce: How Black Women Cope with Their Broken Marriages," *Ebony Magazine* 39 (Nov. 1983):59, 60, 64, 66.

93. Sidney M. De Agnelis, "The Complete Guide to Divorce," *New Woman*, April 1986, pp. 132–39; and Elizabeth Stark, "Friends Through It All" (report of a study conducted by Constance Ahrons at the University of Southern California), *Psychology Today*, May 1986, pp. 54–60.

94. Marie Edwards and Eleanor Hoover, *The Challenge of Being Single* (Los Angeles, 1974), 194; and Lynn Shahan, *Living Alone & Liking It* (Beverly Hills, 1981), 24–25.

95. Sonya Friedman, *Men Are Just Desserts* (New York, 1983); Sonya Friedman, *A Hero Is More Than Just a Sandwich* (New York, 1987); Dan Kiley, *The Peter Pan Syndrome: Men Who Have Never Grown Up* (New York, 1983); Dan Kiley, *The Wendy Dilemma: When Women Stop Mothering Their Men* (New York, 1984); Merle Shain, *Some Men Are More Perfect Than Others* (New York, 1973); Rose DeWolf, *How to Raise Your Man* (New York, 1983); Judi Miller, *How to Ask a Man* (New York, 1978); Penelope Russianoff, *Why Do I Think I Am Nothing Without a Man?* (New York, 1982); Robin Norwood, *Women Who Love Too Much* (New York, 1985); see also Robin Norwood, *Answers from Women Who Love Too Much* (New York, 1988); Susan Forward and Joan Torres, *Men Who Hate Women & The Women Who Love Them* (New York, 1986); and Connell Cowan and Melvyn Kinder, *Smart Women, Foolish Choices* (New York, 1985).

96. Herb Goldberg, *The Hazards of Being Male* (New York, 1976), 159; and *The New Male Female Relationship* (New York, 1983), 15.

97. Warren Farrell, *Why Men Are the Way They Are* (New York, 1986), 136–37, 184–85, 366–68.

98. "When Your Friends Break Up," *Des Moines Register*, Aug. 10, 1986; "Restaurant foots bill on divorce settlement day," *Chicago Tribune* Sept. 3, 1986; and Hallmark Cards, Inc., "It's better to have loved and lost . . .," 1988.

99. Harvey J. Sepler, "Measuring the Effects of No-Fault Divorce Laws Across Fifty States: Quantifying a Zeitgeist," *Family Law Quarterly* 15 (Spring 1981):65–102 and Michael Wheeler, *No-Fault Divorce* (Boston, 1974), 170–85.

## Epilogue

1. Quoted in *Niles' Weekly Register* 28, 717 (June 11, 1825):229.

2. Richard Lewinsohn, *A History of Sexual Customs* (English ed., New York, 1958), 396–98; and Bernard A. Weisberger, "Liberty and Disunion," *American Heritage* 22 (1971):103.

3. James Hitchcock, "American Culture and the Problem of Divorce," *Thought* 58 (March 1983):63, 70.

4. Alan A. Stone, "Emotional Aspects of Contemporary Relations: From Status to Contract," 397–414; Thomas J. Espenshade, "The Recent Decline of American Marriage: Blacks and Whites in Comparative Perspective," 53–90; and Kingsley Davis, "The Future of Marriage," 25–52—all in Kingsley Davis and Amyra Grossbard-Shechtman, *Contemporary Marriage: Comparative Perspectives on a Changing Institution* (New York, 1985).

5. Leonard P. Strickman, "Marriage, Divorce and the Constitution," *Family Law Quarterly* 14 (Winter 1982):310–23. *Boddie v. Boddie* 401 U.S. 371 (1971). Other cases that underwrote freedom of marital choice were *United States v. Kraus* 409 U.S. 434 (1973) and *Zablocki v. Redhail* 434 U.S. 374 (1978).

6. Christopher Lasch, *The World of Nations: Reflections of American History, Politics, and Culture* (New York, 1973), 37–43.

7. For traditional Catholic views of marriage see Fulton J. Sheen, *Three to Get Married* (New York, 1951), 270–86. Changes in Catholic views of divorce are discussed in Arland Thornton, "Changing Attitudes Toward Separation and Divorce: Causes and Consequences," *American Journal of Sociology* 90 (January 1985):856–72. Catholic divorce support groups are described in James J. Young, "The Divorced Catholics Movement," *Journal of Divorce* 2 (Fall 1978):83–97. Annulments are discussed in Daryl Olszewski, *Everyday Theology for Catholic Adults* (Milwaukee, 1989), 87.

8. Margaret Mead, *Male and Female* (New York, 1949), 195.

9. Hayden White, "The Burden of History," *History and Theory* 5 (1966):132.

10. Quotes from Avodah K. Offit, "The New Togetherness," *McCall's Magazine* March 1986, pp. 20, 149. See also Francesca M. Cancian, *Love in America: Gender and Self-Development* (Cambridge, Mass., 1987), 3; and Esther O. Fisher, *Divorce: The New Freedom* (New York, 1974), 1–11. For discussions of education for marriage see Max Rheinstein, "The Law of Divorce and the Problem of Marriage Stability," *Vanderbilt Law Review* 9 (June 1956):633–64; and Doris Jonas Freed and Henry H. Foster, Jr., "Divorce American Style," *Annals of the American Academy of Political and Social Science* 383 (May 1969):70–88. Expectations of marriage are revealed by interviews in Marilyn Little, *Family Breakup* (San Francisco, 1982), 17–114. That traditional views of women and marriage persist is demonstrated in Angela Corencamp, John F. McClymer, Mary M. Moynihan, and Arlene C. Vadum, *Images of Women in American Popular Culture* (New York, 1985), esp. 201–19.

11. Quotes are found in Pat Wingert and Barbara Kantrowitz, "The Day Care Generation," *Newsweek,* special edition on "The 21st Century Family" (Winter/Spring 1990), 86–87. Four years earlier, a *Newsweek* article had asserted that "child care is now an item on the national agenda." "Changes in the Workplace," *Newsweek,* March 31, 1986, p. 57. For discussions of the problem of child care see Annegret Ogden, *The Great American Housewife* (Westport, Conn., 1986), 234–47; and Amy Swerdlow, Renate Bridenthal, Joan Kelley, and Phyllis Vine, *Household and Kin: Families in Flux* (Old Westbury, N.Y., 1981), 131–46.

12. For an elaboration on this point, see Richard Neely, "Barter in the Court," *New Republic,* February 10, 1986, pp. 13–14, 16. See also Judge Neely's *The Divorce Decision* (New York, 1984).

13. The need for more equitable financial arrangements in contemporary divorce is discussed in Heather L. Ross and Isabel V. Sawhill, *Time of Transition: The Growth of Families Headed by Women* (Washington, D.C.,

1975), esp. 173–79; Lynne Carol Halem, *Divorce Reform: Changing Legal and Social Perspectives* (New York, 1980), 287–92; Jean Renvoize, *Going Solo: Single Mothers by Choice* (Boston, 1985), esp. 130–38; and Terry J. Arendell, "Women and the Economics of Divorce in the Contemporary United States," *Signs* 13 (1987):121–35.

14. Lenore J. Weitzman, *The Divorce Revolution: The Unexpected Social and Economic Consequences for Women and Children in America* (New York, 1985), 113–21. See also Frances Leonard, "The Disillusionment of Divorce for Older Women," *Gray Paper* No. 6 (Washington, D.C., Older Women's League, 1980), 8–9; and Henry H. Foster and Doris Jonas Freed, "Commentary on Equitable Distribution," *New York Law School Law Review* 26, 1 (1981):47–48.

15. Many books about divorce include lists of references to resources and organizations. See for example Lucia H. Bequaert, *Single Women: Alone and Together* (Boston, 1976). An argument for tailoring resources to fit a particular group's needs is found in Toni L'Hommedieu, *The Divorce Experience of Working and Middle Class Women* (Ann Arbor, 1984); and Dorothy W. Cantor, "School-based Groups for Children of Divorce," *Journal of Divorce* 1 (Winter 1977):183–263. Examples of divorce workshops are described in David M. Young, "The Divorce Experience Workshop," *Journal of Divorce* 2 (Fall 1978):37–47; and Young, "The Divorced Catholics Movement," 83–97. A report on workshops and other resources for children of divorce was discussed in "Reading, Writing and Divorce," *Newsweek Magazine,* May 13, 1985, p. 74.

16. For a review of the lack of gender awareness in divorce counseling, see Kristina L. Lund, "A Feminist Perspective on Divorce Therapy for Women," *Journal of Divorce* 13, 3 (1990):57–67.

17. Judith Stern Peck, "The Impact of Divorce on Children at Various Stages of the Family Life Cycle," 81–108, and Joseph Guttmann and Marc Broudo, "The Effect of Children's Family Type on Teachers' Stereotypes," 315–28,—both in *Journal of Divorce* 12, 2/3 (1988/89); and Judith S. Wallerstein and Sandra Blakeslee, *Second Chances: Men, Women, and Children a Decade After Divorce* (New York, 1989).

18. See, for example, Robert D. Allers, *Divorce, Children, and the School* (Princeton, N.J., 1982); William F. Hodges, *Interventions for Children of Divorce: Custody, Access, and Psychotherapy* (New York, 1986); and Marla Beth Isaacs, *The Difficult Divorce: Therapy for Children and Families* (New York, 1986).

19. Ernest S. Burch, Jr., "Marriage and Divorce Among the North Alaskan Eskimos," in Paul Bohannan, ed., *Divorce and After* (New York, 1970), 152–81.

20. Among the experts who note that Americans continue to marry despite the high divorce rate are Kingsley Davis, "The Meaning and Significance of Marriage in Contemporary Society," 20–21, and John Modell, "Historical Reflections on American Marriage," 181–96—both in Davis and Grossbard-Schechtman, eds., *Contemporary Marriage.* For other examples of favorable predictions for marriage and the family, see J. P. Lichtenberg, *Divorce: A Social Interpretation* (New York, 1931), 419–59;

Kingsley Davis, "Statistical Perspective on Marriage and Divorce," *Annals of the American Academy of Political and Social Science* 272 (November 1950):9–21; Joseph Epstein, *Divorced in America* (New York, 1974), 317–18; and Amy Swerdlow et al., *Household and Kin*, 38–45.

21. Family types are discussed in Lawrence Stone, "Family History in the 1980s: Past Achievements and Future Trends," *Journal of Interdisciplinary History* 22 (Summer 1981):51–87. Descriptions of the single-parent family are found in "Playing Both Mother and Father," *Newsweek*, July 15, 1985, pp. 42–43; and "The Single Parent: Family Albums," *Newsweek*, July 15, 1985, pp. 44–50. The blended family is described in Barbara Kantrowitz and Pat Wingert, "Step by Step," 24–27, 30, 34; and gay, lesbian, and skip-generation families in Jean Seligmann, "Variations on a Theme," 38–40, 44 46—both in *Newsweek*, special edition on "The 21st Century Family," (Winter/Spring 1990), 24–27, 30, 34.

# Index

Abandonment. *See* Desertion

Absence at sea, 14

Abuse: in the 17th and 18th centuries, 12, 20–21, 29; in the 19th century, 37–38, 55, 61, 74, 81–82, 126; in the 20th century, 133, 180; and African Americans, 133; and bed and board divorce, 12; and cruelty, 81–82; and the debate about divorce, 61, 74; and the divorce rate, 81–82; emotional, 180; and expectations of marriage, 55; and mental instability, 81; and women's rights, 74. *See also* Cruelty; Mental abuse/cruelty; Physical abuse

Adams, Abigail, 55

Adultery: in the 17th and 18th centuries, 10, 12–13, 14, 16, 19–23, 24, 25, 27, 30, 54; in the 19th century [pre-Civil War], 35–38, 40–42, 45–48, 53, 54, 56, 69, 70, 72, 73, 96; in the 19th century [post-Civil War], 71, 76, 82, 93, 98, 102, 112, 126; in the 20th century [early], 117, 124, 125–26, 133, 134, 144, 149, 151; in the 20th century [late], 156, 157, 163, 166, 184; and the acceptance of divorce, 184; and African Americans, 54, 133; biblical views of, 4, 10, 30, 53; and the causal factors of divorce, 149, 151; and collusion, 144, 163; and cruelty, 82; and the debate about divorce, 4, 69, 70, 71, 72, 73, 76; in England, 52; and the English response to colonial divorce, 30; and expectations of marriage, 16, 56; false/sham cases of, 82, 144; and gender, 13, 16, 21, 151; interracial, 35, 36, 37, 54; as the most popular ground for divorce, 124; and

no-fault divorce, 163, 166; proof of, 22; and property issues, 13, 70; and punishment, 25; and the reform of divorce laws, 93, 98, 102, 112; and remarriage, 126; and slaves, 14; testimony about, 22–23; and trials, 21; and the uniform divorce law, 112, 117, 124, 134. *See also name of specific region or state*

Advertising: by attorneys, 165; for remarriage, 172; of runaway spouses, 54–55, 88–89. *See also* Notice of divorce

Advice columnists, 170–71

African Americans: in the 17th and 18th centuries, 22, 28, 54; in the 19th century, 35, 36, 37, 43, 71, 79, 92; in the 20th century [early], 126, 133–34, 152–53; in the 20th century [late], 159–60, 171, 178, 179, 189; and adultery, 54, 133; and the causal factors of divorce, 152–53; clerical divorce of, 54, 71, 79; and debate about divorce, 71; and the divorce rate, 54, 79, 126, 133–34; and ease of divorce, 133; in the future, 189; and grounds for divorce, 133, 159; and the initiation of divorce, 153; post-divorce life of, 178, 179, 189; and property issues, 43; and remarriage, 43, 171; and self-help, 178, 179; and separation, 159; study of divorce among, 159–60. *See also* Interracial adultery; Slaves; *name of specific region or state*

Alabama, 42, 90, 125, 163

Alaska, 163

Alcoholism. *See* Drunkenness

251

65, 66, 77–78, 185; and the divorce process, 187; and the divorce rate, 83; fathers awarded, 83, 92, 125, 168; and guardianships, 39, 40; and the idealization of motherhood, 52; and the initiation of divorce, 153–54; and legislative divorce, 39–40; and no-fault divorce, 162, 163, 165, 166, 167; regional variation in awards of, 92; and the separate spheres concept, 49–53; and the tender years concept, 83, 167; and the uniform divorce law, 121. *See also name of specific region or state*

Child kidnapping, 92

Children: as a continuing issue, 7, 189–90; counseling of, 189–90; divorce statistics about, 119; effect of divorce on, 7, 114, 160–61; fathers as owners of, 83; legal status of, 27, 36–37, 44, 70; and migratory divorce, 70; mulatto, 36–37; need for study of, 189–90; and no-fault divorce, 164, 166–68; and remarriage, 170; and the uniform divorce law, 114. *See also* Blended families; Child care; Child custody; Child support

Child support, 40, 69, 71, 147, 149, 166, 167, 168, 170, 190

Civil War, 78–79

Clergy: in the 19th century, 54, 64, 71, 79, 99, 100, 101, 104, 109–10, 114; in the 20th century, 120, 125, 136; and the debate about divorce, 64; divorce among the, 125; divorce of couples by, 54, 71, 79; and ease of divorce, 136; and the reform of divorce laws, 99, 100, 101, 104; and the uniform divorce law, 109–10, 114, 120. *See also* Religion; *name of specific person*

Coercion into marriage, 57–58

Coercion of a minor [grounds], 41

Co-habitation, 169, 173–74

Colby, Clara, 116

Collusion, 96–97, 143, 144, 149, 155, 163, 164–65

Colorado, 90, 92, 110, 116, 163

Comity principle, 142

Commission on the Status of Women, 175–76

Common-law marriages, 36

Communication, 176–77

Companionate marriage, 55–59

Complex marriage, 68–69

Congress on Uniform Divorce Laws, 123–24

Conjugal relations, 19, 81–82, 126

Connecticut: in the 17th and 18th centuries, 18–22, 32, 53, 56, 57, 60; in the 19th century [pre-Civil War], 43, 44, 46, 47, 52, 60, 62, 68, 72; in the 19th century [post-Civil War], 94–95, 108–9, 110; in the 20th century, 163, 184; African Americans in, 22; alimony in, 47; bed and board divorce in, 19; and the debate about divorce, 60, 62, 68, 72; divorce rate in, 53, 60, 94–95; ease of divorce in, 21–22, 62, 72, 94–95; and expectations of marriage, 56, 57; grounds in, 19, 20–21, 46, 56, 68, 108–9; happiness in, 46, 53; initiation of divorce in, 52; judicial divorce in, 18–19; legislative divorce in, 44; no-fault divorce in, 163; omnibus clause in, 46, 68, 108–9; property issues in, 20, 43, 47; remarriage in, 20, 72; response to divorce in colonial, 32; and the right to divorce, 184; and the separate spheres concept, 52; and the uniform divorce law, 108–9, 110; utopian reformers in, 68

Consanguinity, 4, 25, 41, 45, 60

Constitution, U.S., 111, 117, 120, 134–35, 139–43

Contract: marriage as a, 4, 12, 60, 71–78, 84, 164

Contracts: fraudulent, 19, 46; Jefferson's views of, 31–32; separation, 54

Contracts, marital: in the 17th and 18th centuries, 27–28; in the 19th century, 43–44, 57, 60; in the 20th century, 164, 169, 172; and expectations of marriage, 57; as renewable, 164; and social class, 43; in the South, 27–28, 43–44

Contract theory of divorce, 162, 184

Conviction/commission of a felony, 42, 46, 47, 96, 102, 117, 126, 134, 150

Counseling, 173, 178–79, 180, 186–87, 188, 189–90

Counter-suits, 151

Courts. *See* Ecclesiatical courts; Judicial divorce

Coxe, Margaret, 58

*Crouch v. Crouch*, 142

Cruelty: in the 17th and 18th centuries, 12–13, 14, 21, 25, 26, 28; in the 19th century [pre-Civil War], 35, 37, 40, 41, 42–43, 45–46, 47, 48; in the 19th century [post-Civil War], 79, 81–82, 89–90, 93, 94, 98, 101, 102, 129; in the 20th century [early], 117, 124, 125, 126, 129, 133, 134, 136, 144, 149–50; in the 20th century [late],